797,885 Books
are available to read at

Forgotten Books

www.ForgottenBooks.com

Forgotten Books' App
Available for mobile, tablet & eReader

ISBN 978-1-333-55874-1
PIBN 10519552

This book is a reproduction of an important historical work. Forgotten Books uses state-of-the-art technology to digitally reconstruct the work, preserving the original format whilst repairing imperfections present in the aged copy. In rare cases, an imperfection in the original, such as a blemish or missing page, may be replicated in our edition. We do, however, repair the vast majority of imperfections successfully; any imperfections that remain are intentionally left to preserve the state of such historical works.

Forgotten Books is a registered trademark of FB &c Ltd.
Copyright © 2015 FB &c Ltd.
FB &c Ltd, Dalton House, 60 Windsor Avenue, London, SW19 2RR.
Company number 08720141. Registered in England and Wales.

For support please visit www.forgottenbooks.com

1 MONTH OF FREE READING

at www.ForgottenBooks.com

By purchasing this book you are eligible for one month membership to ForgottenBooks.com, giving you unlimited access to our entire collection of over 700,000 titles via our web site and mobile apps.

To claim your free month visit: www.forgottenbooks.com/free519552

* Offer is valid for 45 days from date of purchase. Terms and conditions apply.

English
Français
Deutsche
Italiano
Español
Português

www.forgottenbooks.com

Mythology Photography **Fiction** Fishing Christianity **Art** Cooking Essays Buddhism Freemasonry Medicine **Biology** Music **Ancient Egypt** Evolution Carpentry Physics Dance Geology **Mathematics** Fitness Shakespeare **Folklore** Yoga Marketing **Confidence** Immortality Biographies Poetry **Psychology** Witchcraft Electronics Chemistry History **Law** Accounting **Philosophy** Anthropology Alchemy Drama Quantum Mechanics Atheism Sexual Health **Ancient History Entrepreneurship** Languages Sport Paleontology Needlework Islam **Metaphysics** Investment Archaeology Parenting Statistics Criminology **Motivational**

Alvan Clark & Sons
Artists in Optics

Deborah Jean Warner

Museum of History and Technology

Smithsonian Institution

SMITHSONIAN INSTITUTION PRESS

CITY OF WASHINGTON • 1968

Publications of the United States National Museum

The scholarly and scientific publications of the United States National Museum include two series, *Proceedings of the United States National Museum* and *United States National Museum Bulletin*.

In these series, the Museum publishes original articles and monographs dealing with the collections and the work of its constituent museums—the Museum of Natural History and the Museum of History and Technology—setting forth newly acquired facts in the fields of anthropology, biology, history, geology, and technology. Copies of each publication are distributed to libraries, to cultural and scientific organizations, and to specialists and others interested in the different subjects.

The *Proceedings*, begun in 1878, are intended for the publication, in separate form, of shorter papers from the Museum of National History. These are gathered in volumes, octavo in size, with the publication date of each paper recorded in the table of contents of the volume.

In the *Bulletin* series, the first of which was issued in 1875, appear longer, separate publications consisting of monographs (occasionally in several parts) and volumes in which are collected works on related subjects. *Bulletins* are either octavo or quarto in size, depending on the needs of the presentation. Since 1902 papers relating to the botanical collections of the Museum of Natural History have been published in the *Bulletin* series under the heading *Contributions from the United States National Herbarium*, and since 1959, in *Bulletins* titled "Contributions from the Museum of History and Technology," have been gathered shorter papers relating to the collections and research of that Museum.

This work forms volume 274 of the *Bulletin* series.

FRANK A. TAYLOR
Director, United States National Museum

Table of Contents

Introduction
Part I
    Biographical Outline of Alvan Clark and His Sons   3
Part II
    Catalog of Astronomical Instruments Made and Remade
    by Alvan Clark & Sons, 1844–1897   39
Appendix
    Paintings by Alvan Clark   113

FIGURE 1.—Alvan Clark (center) with his sons Alvan Graham Clark (left) and George Bassett Clark (right). Courtesy Lick Observatory.

Introduction

Three instrument makers—Alvan Clark and his sons, George Bassett and Alvan Graham—figured importantly in the great expansion of astronomical facilities which occurred during the second half of the 19th century. Almost every American observatory built during this period, and some observatories abroad, housed an equatorial refracting telescope, and often auxiliary apparatus as well, made by the Clarks. Five times the Clarks made the objectives for the largest refracting telescopes in the world; and the fifth of their efforts, their 40-inch lens at the Yerkes Observatory, has never been surpassed. Their optical work, which was recognized as unexcelled anywhere in the world, was the first significant American contribution to astronomical instrument making. American telescopes had been made before, but none compared to those of European manufacture; by the end of the 19th century, however, partly because of the example set by Alvan Clark & Sons, several other Americans were making fine astronomical instruments.

Fortunately, the Clarks lived at a time when men could afford as well as appreciate their work. Astronomy had caught the public imagination and the private purse. Astronomers, it was thought, could penetrate far into space, discover new worlds, and evidence the glory of God. Equally important, telescopes were obvious symbols of their donor's opulence; these well-publicized monuments were seldom compared qualitatively, but were always described by their size. Thus, when rich Americans wanted to express their love for learning, and also wanted to insure their fond remembrance, they often endowed telescopes and observatories.

Many historical accounts discuss the work of Alvan Clark but neglect the important contributions made by his two sons. There are several obvious explanations for this. Through a common contraction Alvan Clark & Sons became Alvan Clark. Before turning to astronomical instruments, Alvan Clark was noted for his work in other fields; George Bassett and Alvan Graham, on the other hand, devoted their entire professional careers to Alvan Clark & Sons. Another reason is that toward the end of his life Alvan Clark wrote a frequently reprinted and often quoted autobiography; but his sons were more modest. Although each had his own specialty—Alvan and Alvan Graham did optical work while George did mechanical work—it is impossible, in almost all instances, to identify the man most responsible for a particular job. All three Clarks should be remembered equally for the achievements of Alvan Clark & Sons.

No discussion of Alvan Clark & Sons would be complete without a list of the instruments made in their shop. Previous articles on the Clarks have emphasized, usually to the exclusion of all else, their objectives over $18\frac{1}{2}$ inches in diameter. These were indeed great achievements. The Clarks, however, should also be remembered as very prolific craftsmen who made a wide variety of astronomical instruments, including equatorial refractors, spectroscopes, chronographs, and micrometers.

The second half of this volume is a descriptive catalog of instruments made and remade by the Clarks between the dinner bell experiment of 1844 and the death of Alvan Graham Clark in 1897.

Part I

Biographical Outline of Alvan Clark and His Sons

Alvan Clark was a characteristic New England Yankee.[1] To all his pursuits he brought common sense, perseverance, and a desire for perfection. From childhood he was interested in the mechanical workings of things. His visual perception, which was unusually keen, enabled him to align rifle sights precisely, paint striking likenesses of people, and detect minute errors in the figure of a lens. Alvan Clark's personal habits were plain. Always God-fearing, he was never a church member. He voted Republican but was disinterested in politics. And although friendly to those who sought him out, his efforts to seek society were reserved.[2]

Clark was born in Ashfield, Massachusetts, in 1804, the fifth of ten children of Abram and Mary Bassett Clark. Little is known of Abram Clark other than that he was descended from a Mayflower passenger, Thomas Clark; that he owned and operated a rocky farm, a sawmill, and a gristmill in Ashfield; and that he left Alvan a patrimony of fifty dollars. Alvan received his formal education, such as it was, at a small grammar school located on the family farm. He was described as "a dreamy, absent-minded boy, not showing any particular talent."[3] At

---

[1] Autobiography of Alvan Clark published, among other places, in *Sidereal Messenger*, vol. 8 (1889), pp. 109–117. Unless otherwise noted, most biographical data are from this.

[2] It was frequently noted that Cambridge residents seldom knew the Clarks, or even where the Clark workshop was located. See Charles Palmer, "Two Hours with Alvan Clark, Sr.," *Popular Astronomy*, vol. 35 (1927), pp. 143–145.

[3] Frederick G. Howes, *History of the Town of Ashfield* (Ashfield, Mass., n.d.), p. 327.

FIGURE 2.—Alvan Clark.

the age of seventeen Alvan began working with his older brother Barnabus in a wagon maker's shop.

During this time Alvan visited Hartford and there had his first exposure to art; he was so inspired by the experience that he quit the shop and began to learn drawing and engraving. Two years later he felt proficient enough to carry a portfolio of his work to Boston, where he spent the winter of 1824. The following summer he returned to Ashfield and traveled through the Connecticut Valley, painting small portraits in ink and watercolor. It is more than likely that he met there several people who later were involved with astronomy and Clark instruments. He probably called on the family of Edward Hitchcock, since his future bride, Maria Pease, was boarding with them at that time. Hitchcock, then pastor of the Congregational Church in Conway, Massachusetts, was a vocal amateur astronomer. Under his presidency Amherst College acquired an early Clark telescope. The nearby Sanderson Academy in Ashfield attracted Elijah Burritt, author of the popular *Geography of the Heavens*, and Mary Lyon, who went on to found Mount Holyoke Seminary and there instituted astronomy in the first course of study. A Clark telescope was added to the Mount Holyoke observatory in 1880.

New England textile towns were then just getting under way and attracting young people from the surrounding countryside. In the autumn of 1825 Alvan Clark answered an advertisement in a Boston paper and was hired by Mason & Baldwin, subcontractors to the Merrimac Manufacturing Company of East Chelmsford. He worked nine hours a day in winter, ten in summer, and earned eight dollars a week while he learned the art of engraving the mills and cylinders used to print calico patterns. In his spare time he could attend the popular astronomy lectures given by Warren Colburn, superintendent of the Merrimac Company.

In the first wedding ceremony performed in the town of Lowell, as the incorporated East Chelmsford was called, Alvan Clark married Maria Pease on 25 March 1826. They soon became the parents of four children: Maria Louisa and Caroline Amelia, as well as George Bassett and Alvan Graham. The Clarks lived to celebrate their sixtieth wedding anniversary, an event noted by *Science* magazine.[4]

In 1826 working conditions in the engraving shop were far from harmonious: problems arose when the English-born master engravers jealously guarded their techniques from the American employees; also, David Mason strenuously disagreed with Matthias W. Baldwin over Baldwin's steam engine experiments. Clark was glad to be placed in charge of a branch shop opened in Providence, Rhode Island, in 1827, and to move to New York City to open another shop the following year. In the spring of 1832 he accepted a position with Andrew Robeson's print works in Fall River, Massachusetts.

While working as an engraver Alvan Clark continued to paint as an avocation, and during his four years in New York he found excellent

[4] *Science*, vol. 7 (1886), pp. 303-304.

Figure 3.—Maria Pease (Mrs. Alvan) Clark, painted by Alvan Clark. Oval miniature owned by Theodore C. Hollander, grandson of Alvan Graham Clark.

FIGURE 4.—Abram Clark, painted by his son Alvan Clark. This miniature portrait owned by Mrs. Albert W. Rice is in the Worcester Art Museum, Worcester, Massachusetts.

opportunities to study art. Although he was not a member, he exhibited at the National Academy of Design as early as 1829.[5] This academy had recently been founded by Samuel F. B. Morse as a protest by the younger artists against the established, and exclusive, American Academy of Arts. In 1830, and for many years thereafter, Clark also exhibited at the Athenaeum Gallery in Boston,[6] and his pictures were often shown at

---

[5] National Academy of Design Exhibition Record, 1826–1860, vol. 1. Also, private correspondence with Alice G. Melrose, assistant to the director of N.A.D.

[6] Mabel Swan, *The Athenaeum Gallery, 1827–1873* (Boston, 1940), pp. 181, 212.

the smaller private Boston art exhibitions. His work, which seems to have been exclusively portraits, shows no influence of the more imaginative schools of painting of that time; it was said, however, to be "characterized by a rare fidelity and accuracy, and by a rugged underlying strength on the part of the artist."[7] The faithfulness of Clark's portraits was sometimes obtained by the use of a prism—a camera lucida—to outline distinguishing features.[8]

In 1836 Alvan Clark renounced engraving to earn his living by painting portraits and miniatures. He attributed this decision to the encouragement and example of Lucius Manlius Sargent, an itinerant temperance lecturer. During his tour through Fall River, Sargent sat for an ivory miniature by Clark, for which he paid forty dollars—twice as much as Clark had ever received for a single portrait.[9] Clark thereupon moved his family to Cambridgeport (as the center section of Cambridge was then called) and opened a studio in the artists' district of Boston. Years later he recalled, "In that room on Tremont Street I painted heads for twenty years, and took in over $20,000, making a living and laying up a little something besides."[10] Portraits of a president of Harvard, Thomas Hill, and the statesman Daniel Webster hung in the Clark home. In later years criminals were deterred from tampering with the Clark telescope factory by a portrait of Constable Clapp, the renowned rogue catcher. Desiring a likeness of his astronomical correspondent, W. R. Dawes, before he was able actually to meet him, Clark painted his portrait from a daguerreotype. He also painted portraits of Nathan Loomis and the chemist Robert Hare. Clark had met Hare in 1856 and had requested an opportunity to paint his portrait. A Clark family favorite, this picture hung in their home until 1869, when it was sold to the Smithsonian Institution for $100.[11] Although he evidently enjoyed portraiture, Clark readily renounced it in favor of telescope construction, a move that may well have been encouraged by the strong competition from the popular portrait photographers. He kept his studio open until 1860, however, when the

---

[7] Garth Galbraith, "The American Telescope Makers," *The Cambridge Chronicle* 12 March 1887.

[8] Samuel L. Gerry, "The Old Masters of Boston," *New England Magazine*, n.s. vol. 3 (1891), pp. 686–687.

[9] Alvan Clark autobiography, op. cit., p. 112.

[10] Quoted in Garth Galbraith, "The American Telescope Makers," op. cit.

[11] Alvan Clark to Joseph Henry, 5 October 1869 (letter in Smithsonian Institution Archives).

FIGURE 5.—Alvan Clark as an artist, painted by George Hollingsworth. This portrait, formerly owned by the Clarks, now hangs in the Harvard College Observatory. Courtesy Fogg Art Museum, Cambridge, Massachusetts.

Alvan Clark & Sons telescope business appeared lucrative enough to support his family.

Alvan Clark was widely known as a sharpshooter as well as a portrait painter, and there are numerous legends of his marksmanship. With rifles and bullets of his own make he was able to "put bullet after bullet

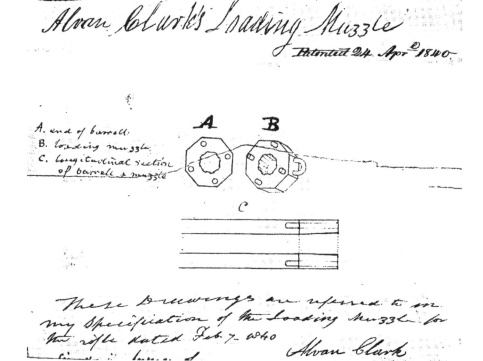

FIGURE 6.—False loading muzzle for rifles, patented by Alvan Clark on 24 April 1840. Drawing from patent papers.

through a distant board with such precision that one would say only a single shot had been fired."[12] Clark reputedly said to Joseph Henry, "You say . 'every man ought to make himself a master in some one thing.' Well, I think I am the best rifle shot in the world."[13]

Dissatisfied with the accuracy of common muzzle-loading rifles, Clark devised a false loading muzzle. Its purpose was to secure the patch from injury, to facilitate as tight a loading as could be wished, and to achieve as perfect a delivery as possible.[14] This invention, patented in 1840, consisted of a hollow cylinder which is fastened to the end of the muzzle

---

[12] William B. Hawkins, article in a Boston newspaper ca. 1893, reprinted *Popular Astronomy*, vol. 34 (1926), p. 379.

[13] Quoted in Garth Galbraith, "The American Telescope Makers," op. cit.

[14] Alvan Clark, "On Rifle Shooting," *The American Repertory*, vol. 3 (1841), pp. 164–169.

FIGURE 7.—Mrs. Charles Henry Cummings, painted by Alvan Clark. Courtesy Museum of Fine Arts, Boston, Massachusetts.

during loading.[15] At long range rifles fitted with this device were more accurate than any others of that period.[16] Rifles equipped with Clark's false muzzle were made exclusively by Edwin Wesson, who paid him two dollars for each such rifle made. Wesson later sold to other riflemakers the right to make these instruments, at a cost of three dollars each, two of which he passed on to Clark.[17]

---

[15] U.S. Patent 1565 (24 April 1840).
[16] "The Muzzles of Rifles and Rifled Cannon," *Scientific American*, n.s. vol. 4 (1861), p. 337.
[17] Ned H. Roberts, *The Muzzle-loading Cap Lock Rifle* (Manchester, N.H., 1940), pp. 99–101.

As has often been related, Alvan Clark became a telescope maker almost by accident. Interest in astronomy became widespread in 1844, spurred by the appearance of the great comet in the previous year. Wealthy Bostonians raised money for a German equatorial refractor, equal to the largest in the world, for Harvard College. During that same year the dinner bell broke at Phillips Academy at Andover, where George Bassett Clark was enrolled as a student in preparation for entering Harvard. Following Newton's example, George melted down this bell metal to make a reflecting telescope. Alvan watched his son's experiment with growing enthusiasm and, like any father, could not refrain from giving him the "benefit" of his "maturer judgment"; [18] he then promptly became involved with the construction of telescopes. The record intimates, however, that neither of his sons worked with him much before 1850.

The firm of Alvan Clark & Sons grew out of a small shop in East Cambridge in which George Bassett Clark made and repaired scientific instruments. The Clarks themselves do not seem to have publicly recorded the establishment of their company. The date 1850—given in 20th-century advertisements [19]—must refer to the start of George's shop. For Alvan Graham was still serving a mechanical apprenticeship in 1850; and until 1860, when he finally closed his Boston portrait studio, Alvan Clark could spend only his spare time working on astronomical instruments.[20]

Although George was directly responsible for the first telescope and the nucleus of the company, we know less about him than about his father or brother.[21] This is perhaps because his constant devotion to the business kept him from pursuing other activities. He was born in Lowell in 1827, and prior to his two years at Andover he attended a grammar school, a high school, and Mr. Whitman's private school in Cambridge. George never made it to Harvard. After graduation from Andover he was attracted by the railroads, which had just begun to spread across New England, and spent a couple of years as a civil engineer on the Boston and Maine and the Ogdensburg and Lake Champlain lines. In 1848 the gold rush lured him to California: he returned East within a year,

---

[18] Alvan Clark autobiography, op. cit., p. 113.

[19] See advertisements in *Popular Astronomy*, vol. 13 (1905).

[20] "The Alvan Clark Establishment," *Scientific American*, vol. 57 (1887), pp. 198–199.

[21] Most available biographical information on George Bassett Clark is from his obituary in *Proc., American Academy of Arts and Sciences*, vol. 27 (1891–1892), pp. 360–363.

"richer in experience than in worldly goods," and reluctant to discuss his adventures.

From his youth, when he used a lathe to make toys for his friends, George had always been interested in the "mechanical arts." Once the firm was established this interest was pursued to the exclusion of almost all else. The Clarks never had a line of goods, but made most of their instruments to order. While some customers knew just what kind of instruments they wanted, others had only rough ideas of the research they wished to follow. George, therefore, was frequently called upon to design as well as to construct a variety of scientific apparatus. He worked indefatigably on each instrument, both in the shop and after delivery, until he was satisfied that it performed as well as possible. He apparently

FIGURE 8.—Charles Henry Cummings, painted by Alvan Clark. Courtesy Museum of Fine Arts, Boston, Massachusetts.

Figure 9.—Robert Hare, painted by Alvan Clark. This portrait of the great American chemist was purchased by the Smithsonian Institution in 1869. Courtesy National Portrait Gallery, Smithsonian Institution.

allowed himself few holidays and was loathe to rest until the orders at hand were completed. That his efforts were appreciated is obvious from comments dropped, seemingly incidentally, by people who relied on his craftsmanship: while describing their new Clark instruments astronomers often singled out George's untiring skill and perseverance for special praise.[22] In 1878 George was elected a fellow of the American Academy of Arts and Sciences; in 1882, and for several years thereafter, he was a member of its Rumford Committee.

---

[22] Edward C. Pickering, "Henry Draper Memorial," *Scientific American Supplement*, vol. 24 (1887), p. 9604.

Alvan Graham Clark was as deeply involved in the family business as was George. In talent and temperament, however, the brothers differed considerably.[23] While George did mechanical work, Alvan Graham, with an eye as keen as his father's, figured and tested the object glasses. George was little known outside his work; Alvan Graham is described as unusually attractive in both social intercourse and personal appearance. He was fond of the companionship of intelligent men, poets as well as scientists, personal friends as well as casual visitors. With a love for literature, and a remarkably retentive memory, Alvan Graham could, and apparently did, quote from the poets "almost indefinitely." His sociability was doubtless enhanced by his wife, Mary Mitchell Willard, a member of a large and influential Cambridge-Harvard family.

Alvan Graham was born in Fall River in 1832. As a schoolboy in East Cambridge, at the time of his father's first telescope experiments, he wrote prize essays on the casting and grinding of mirrors.[24] At age sixteen he entered a machine shop, where he spent four or five years learning the machinist's trade before entering the family firm. While testing object glasses Alvan Graham discovered a number of interesting and difficult double stars; the most famous, the companion of Sirius, earned him the 1862 Lalande Prize of the Paris Académie des Sciences.[25] He joined the American Association for the Advancement of Science in 1879 and the following year was elected to fellowship; in 1881 he became a resident fellow of the American Academy of Arts and Sciences; in 1894 he was elected a member of the Société Astronomique de France.[26]

The first astronomical instruments Alvan Clark worked with were metal reflectors, as they were reputedly easier than refractors. He seems to have made a number of these telescopes, with apertures as large as 8 inches. During the winter of 1847–48, using a freshly polished 7½-inch speculum, he made a diagram of the stars in the Orion Nebula. To guarantee the honesty of his observations, he refrained from studying previous maps until he had drawn his own.[27] William Cranch Bond,

---

[23] Most available biographical information on Alvan Graham Clark is in his obituary in Proc., *American Academy of Arts and Sciences*, vol. 33 (1897–1898), pp. 520–524.

[24] "Alvan Graham Clark," *Dictionary of American Biography* (New York, 1930), vol. 4, p. 120.

[25] *Comptes Rendus, Académie des Sciences*, vol. 55 (1862), pp. 936–937.

[26] *Bulletin, Société Astronomique de France*, vol. 11 (1897), p. 300.

[27] Alvan Clark, "Telescopes," *The Boston Courier*, 13 November 1848.

FIGURE 10.—George Bassett Clark. Courtesy Lick Observatory.

director of the new Harvard College Observatory, was greatly impressed with Clark's map and pointed out that he had plotted a star which had escaped William Herschel with his 20-foot reflector.[28] Nevertheless, Clark was dissatisfied with the definition and light-gathering power of his mirrors and, around 1846, he began to figure lenses. Within a year he had acquired enough knowledge of optics to perceive and locate, at a glance, certain slight errors of figure in the 15-inch lens of the Harvard telescope. This lens, made by the German firm Merz und Mahler, had cost $12,000. The errors and the cost gave Clark the "hope and courage" he needed to begin making refracting telescopes for sale.[29]

---

[28] Simon Newcomb, "The Story of a Telescope," *Scribner's Monthly*, vol. 7 (1873–1874), p. 44.

[29] Alvan Clark autobiography, op. cit., p. 113.

FIGURE 11.—Alvan Graham Clark. Courtesy Lick Observatory.

From the first, Clark lenses were probably equal to any ever made,[30] and undoubtedly better than those made by the only other serious and contemporary American lens maker, Henry Fitz.[31] Fitz was immediately successful, however, while Clark's talents were little appreciated until William R. Dawes, the well-known British double-star observer, published reports of the extraordinary performance of his lenses. Alvan Clark himself attributed the increasing number of orders to Dawes' publicity. Simon Newcomb may have been exaggerating, but he certainly expressed a grain of truth when he wrote that had Alvan Clark "been a citizen of any other civilized country, [he] would have found no difficulty in

---

[30] Simon Newcomb, *Popular Astronomy* (New York, 1878), p. 139.

[31] The Clarks refigured and greatly improved several large Fitz lenses. See below, catalog of Clark instruments.

establishing a reputation. But he had to struggle 10 years with that neglect and incredulity which is the common lot of native genius in this country." [32] Maria Mitchell explained that "Mr. Clark's lack of mathematical learning, or learning of any kind, kept him out of the confidence of the scholarly persons around Boston. His work was too much like empiricism; his claims seemed to be unreliable." [33] But Clark was no less educated nor more empirical than was Fitz. The contrast between Clark's obscurity and Fitz's early recognition can, I think, be accounted for largely by the company the two men kept. Essentially an artist, Clark was unable to fit into the conservative and highly mathematical Cambridge astronomical community; nor did he try to work with the photographers in Boston. Fitz, on the other hand, was called upon to supply optical instruments for the many New Yorkers investigating the new techniques of photography and spectroscopy and their applications to astronomy; as an active member of the American Photographical Society, Fitz met frequently with other members, such as Lewis Morris Rutherfurd, and John William and Henry Draper.

Much of Alvan Clark's obscurity was perhaps his own fault. He was extremely averse to advertising his work in any way. He never published a price list, although astronomers requested it; nor would he show his instruments at any of the popular international exhibitions.[34] What is more, before his correspondence with Dawes, Clark probably did not know how fine his lenses actually were. For a critical evaluation of his early work Clark had turned to the astronomers at Harvard, William Cranch Bond and his son, George Phillips Bond. The Bonds, unfortunately, were jealous of their infrequent and valuable observing time, often in poor health, and constantly besieged by the Boston public who had paid for their new telescope; understandably, the imperfections in Clark's first mirrors quickly eroded their patience.[35] The first lens Clark showed to the Bonds seemed equally unsatisfactory; although the imperfection—the 4-inch achromat showed star images with tails—was later

---

[32] Simon Newcomb, *Popular Astronomy*, p. 139.

[33] Maria Mitchell, Alvan Clark and Telescope Making (ms of lecture, in library of the Maria Mitchell Association, Nantucket, Mass.).

[34] "Alvan Clark," *Harper's Weekly* (1887), p. 631.

[35] See William C. Bond, Diary, 1846–1849, entry for 26 April 1846 (in Bond Papers, Harvard University Archives).

attributed to irregular refraction within the temporary tube, for several years the Bonds were reluctant to recommend Clark's work.[36]

The first recorded sale of a Clark telescope—a 5-inch aperture achromatic refractor—was to William Harvey Wells. As he had taught English and science at Phillips Academy between 1836 and 1847, Wells was very probably the man most responsible for the first dinner bell-reflector experiments. On 4 October 1848 the new refracting telescope was set up at the Putnam Free School in Newburyport, Massachusetts, where Wells was then principal. Simon Newcomb's cynical remark to the contrary, laudatory descriptions of this instrument were widely publicized. The *Boston Courier* of 13 November 1848 carried a lengthy article by Alvan Clark reporting the trials at Newburyport: Wells, although unfamiliar with the positions of stars, could distinguish the close pair in the triple star γ Andromeda, and locate the fifth, and at intervals the sixth, stars in the Orion trapezium.[37] In an uncommon burst of self-advertisement, Clark extensively circulated copies of this notice.[38] Two years later, in his popular *Recent Progress of Astronomy*, Elias Loomis repeated these assertions; and he quoted Charles A. Young to the effect that a similar Clark objective was "one of great excellence, indicating a high degree of finish in respect to the correction of both the chromatic and spherical aberrations."[39] Owing to this publicity Clark apparently sold some other telescopes at this time, but the purchasers are unknown.[40]

Figuring an achromatic lens was difficult enough, but, for the Clarks, obtaining flint and crown glass discs of suitable purity was even harder. The few firms which made good optical glass were in Germany and France, and they kept their techniques secret. Alvan Clark figured some discs of American glass but was soon ready to renounce his efforts until he could find better material.[41] This lack of a local optical glass factory must always have been a nuisance; in 1879, and perhaps at other times

---

[36] H. S. Leavitt, "Clarks' Observatory," in E. M. H. Merrill (ed.), *Cambridge Sketches* (Boston, 1896), p. 150. See also Simon Newcomb, "The Story of a Telescope," op. cit., p. 44.

[37] Alvan Clark, "Telescopes," *The Boston Courier*, 13 November 1848.

[38] Alvan Clark to William C. Bond, 30 March 1849 (letter in Bond Papers, Harvard University Archives).

[39] Elias Loomis, *Recent Progress of Astronomy* (New York, 1850), pp. 252–253.

[40] Alvan Clark autobiography, op. cit., p. 113. See also Simon Newcomb, "The Story of a Telescope," op. cit., p. 45.

[41] Elias Loomis, *Recent Progress of Astronomy*, p. 252.

as well, the Clarks made experimental castings at a Cambridge glassworks.[42] Although the British opticians and scientists had long been experimenting with optical glass, theirs was as irregular as the American product. For these reasons, Alvan Clark made his first lenses from objectives of old instruments.[43] Fortunately, his need for better glass coincided with the midcentury political upheavals that caused George Bontemps, who was privy to the technical secret of making fine optical glass, to emigrate from Paris and to join the Chance Bros. glassmaking company in Birmingham, England. Soon thereafter Chance Bros. began supplying the Clarks with optical discs. Around 1875 the Clarks turned to the house of Feil-Mantois in Paris for their glass. All the large objectives figured by the Clarks were made possible only by the successful castings of these two factories.

Optical blanks were as expensive as they were difficult to obtain. A duty of 30 percent was levied on the purchase price, rather than the actual value, of each imported blank. As these blanks often turned out to be optically worthless, and as they seem to have been nonreturnable, the duty soon became oppressive. Clark finally brought suit against the customs collector; he won the case, but in this minor Bleak House the damages paid only the court costs.[44] In later years the Clarks were able to avoid the duty when the glass was destined for a school, for the U.S. Government, or—as in the case of the 30-inch discs for the Russian observatory at Pulkowa—when imported by a foreign minister.[45]

In 1860, after the order for the 18½-inch lens had been received (see below), the Clarks moved from Prospect Street to larger premises by the Charles River (see fig. 12, p. 41). On an acre and a half on Henry Street, near the Brookline Bridge, they built their workshop, an observatory, and separate dwellings for the Alvan Clarks, the George Clarks, and the Alvan Graham Clarks. Their location could be spotted from afar by the tall telescope tube and mount which was used to test new objectives. The outside grounds were covered with grass and beds of flowers.[46] The factory itself was an unpretentious two-story brick structure, about forty

---

[42] Charles Plumb to E. Mathews, 3 June 1879 (letter in Lick Observatory Archives).

[43] Alvan Clark autobiography, op. cit., p. 113.

[44] Ibid., p. 114. See also Simon Newcomb, "The Story of a Telescope," op. cit., p. 46.

[45] "Clark's Telescope Works," *The Cambridge Chronicle*, 2 January 1892.

[46] "The Alvan Clark Establishment," *Scientific American*, vol. 57 (1887), p. 198.

feet long by twenty-five feet wide, with an ell of the same width and thirty feet long.[47] Under the workshop the Clarks had a long chamber dug for testing optical systems. A fireproof safe, with telegraphic alarms connected with Alvan Clark's bedroom, housed the valuable lenses at night; the lenses actually rested on a railway car which ran in and out of the safe.[48]

In the years before the Civil War the Clarks made and sold about a dozen medium-size lenses; the largest, of 12 inches aperture, showed Mimas, the innermost satellite of Saturn.[49] For many of these they also provided equatorial mounts. Then in 1860 the Clarks received an order for a lens larger than any ever made. The 18½-inch for the University of Mississippi was to have an aperture 3½ inches larger than did the Harvard and Pulkowa instruments. This was the first of the five times that the Clarks would progressively surpass the world's record for existing lens size. In 1873 they finished the 26-inch for the United States Naval Observatory; and twelve years later a similar telescope was erected at the University of Virginia. In 1883 they finished the 30-inch lens for the Russian Observatory at Pulkowa. The 36-inch lens for the Lick Observatory was sent to California in 1887. And the 40-inch objective, mounted at the Yerkes Observatory in 1897, is still the largest lens ever used.

Alvan Clark formally introduced himself to the American scientific community at the 1850 meeting of the American Association for the Advancement of Science in New Haven. He was then elected to membership and attended meetings until that body dissolved, temporarily, ten years later. The 1856 meeting, held at Albany to celebrate the dedication of the Dudley Observatory, was the largest and most representative scientific meeting held in America before the Civil War.[50] Here Alvan Clark met Robert Hare, whose portrait he painted soon thereafter; he very likely also met F. A. P. Barnard, the newly elected president of the University of Mississippi, who four years later commissioned him to build the world's largest telescope. At this meeting Alvan Clark read a paper

---

[47] An occasional correspondent of the Tribune, "Two Giant Telescopes," Boston 5 February [n.y.] (clipping in library of the Maria Mitchell Association).
[48] *Boston Journal of Chemistry*, vol. 7 (1872), p. 56.
[49] *American Journal of Science*, vol. 29 (1860), p. 449.
[50] W. H. Hale, "Early Years of the American Association," *Popular Science Monthly*, vol. 49 (1896), p. 503.

"On a New Method of Measuring Celestial Arcs,"[51] which the extremely practical *Scientific American* hailed as one of the most valuable papers presented that year.

The new double eyepiece micrometer described in the paper was designed to measure celestial distances too great to be brought within the field of view of a single eyepiece powerful enough to see the objects. The two eyepieces moved independently so that the crosshairs of each could be aligned with separate objects. The distance between the eyepieces was then measured, not with a fixed rule or screw, but with calibrated parallel lines ruled on a rectangular glass plate. This plate was placed under the eyepieces of the micrometer so that the ruled lines were aligned with the crosshairs and the celestial objects. If the lines were not the proper distance apart, another plate was ruled by an assistant using a dividing machine made for this purpose. The micrometer was attached to an equatorial refractor of 6 inches aperture, which was made, of course, by the Clarks. To keep the stellar images on the crosshairs a reliable driving clock was necessary; for this Clark used a Bond spring governor. Shortly after George Phillips Bond, in 1851, had described the spring governor,[52] which had been devised for imparting equable motion to a chronograph, the Clarks followed his suggestion and applied the new regulator to telescope drives. At the AAAS meeting of 1856 Benjamin Peirce testified to the regularity of this drive mechanism, which he found "vastly superior in convenience and value" to that of the Harvard telescope.[53] Using this instrument the Clarks measured distances up to one hundred minutes of arc with acceptable accuracy.

Although his sons made frequent trips abroad to obtain glass and to examine telescopes, Alvan Clark went to Europe only once. In 1859 he spent a month visiting his close friend William Dawes. By this time Clark had sold Dawes five telescopes, and their correspondence had been more extensive and their dealings more lucrative than any Clark had had.[54] In company with Dawes, Clark attended a visitation at the Green

---

[51] Alvan Clark, "On a New Method of Measuring Celestial Arcs," *Proc., American Association for the Advancement of Science*, vol. 10 (1856), pp. 108-111.

[52] George P. Bond, "Description of the Apparatus for Observing Transits by Means of a Galvanic Current, Now Used at the Observatory of Cambridge. U.S.," *Monthly Notices, Royal Astronomical Society*, vol. 11 (1851), pp. 163-165.

[53] Quoted in *Scientific American*, vol. 12 (1856), p. 3.

[54] Alvan Clark autobiography, op. cit., p. 114.

wich Observatory, and the June meeting of the Royal Astronomical Society, where he met Sir John Herschel and Lord Rosse. At the R.A.S. meeting Alvan Clark exhibited an improved double eyepiece micrometer wherein the distance between the two separate eyelenses was measured by movable, parallel spider lines mounted in front of the eyepieces.[55] One evening, while using the 8-inch-aperture equatorial refractor he had just brought to Dawes, Clark discovered the duplicity of 99 Herculis. In reply to Mrs. Dawes, who asked what made him think this star had not yet been charted, Alvan Clark said, "I thought it too fine for anybody else." [56]

During his lifetime, undoubtedly in recognition of work done by Alvan Clark & Sons, Alvan Clark received four honorary Master of Arts degrees and two special medals.[57] The first degree came from Amherst College in 1854 when the Clarks made a 7½-inch refractor for that school. Princeton conferred a similar degree in 1865. The old University of Chicago gave him an honorary degree in 1866, the year the 18½-inch was installed in Chicago. Alvan Clark was for many years on the visiting committee of the Harvard College Observatory; in 1874 Harvard made him a Master, with the words, *Artificem egregium, speculatorem rerum coelestium callidum*.[58] It is related that when Dom Pedro, the Brazilian emperor, was in Cambridge in 1876, he wished to see only three persons: Henry Wadsworth Longfellow, Louis Agassiz, and Alvan Clark.[59] On the suggestion of Otto Struve, Czar Alexander III of Russia in 1885 granted Alvan Clark a medal "in acknowledgment of the excellent performance of the great [Pulkowa] object-glass." This medal was of solid gold, $\frac{3}{16}$-inch thick and $3\frac{5}{8}$ inches in diameter.[60] Alvan Clark's other great honor, which came in 1867, was the rarely

---

[55] "Mr. Alvan Clark's New Micrometer for Measuring Large Distances," *Monthly Notices, Royal Astronomical Society*, vol. 19 (1859), p. 324.
[56] Quoted in Garth Galbraith. "The American Telescope Makers," op. cit.
[57] Alvan Clark autobiography, op. cit., pp. 116–117.
[58] Quoted in private correspondence with Kimball C. Elkins of the Harvard University Archives.
[59] "Alvan Clark," *National Cyclopaedia of American Biography* (New York, 1929), vol. 6, p. 440.
[60] Otto Struve to Alvan Clark & Sons, 23 July 1885, quoted in *Appleton's Annual Cyclopaedia*, vol. 10 (1885), pp. 54–55. See also *Scientific American*, vol. 53 (1885), p. 304, and *Science*, vol. 7 (1886), p. 350.

given Rumford Medal of the American Academy of Arts and Sciences.[61]

The Rumford Medal was awarded to Clark in general recognition of his ability to figure nearly perfect lenses, and in particular for his method of local correction. Several stories, indicative of Clark's personality, were told about this prize. Anyone who was being considered by the Rumford Committee was required to submit an essay describing his work. According to one source, Alvan Clark was unwilling to waste time preparing this essay. Instead, he invited the committee to his workshop where he gave them a demonstration of his methods.[62] According to Alvan Clark himself, the Rumford Committee asked him to write on his original methods; he then invited the committee to his shop because since he knew so little of the way others had made lenses, he could not say which of his devices were unique with him.[63] In much the same vein, around 1847 Alvan Clark told Dr. Jacob Bigelow, who had recently been in Munich where the Harvard equatorial had been made, that he was interested in making astronomical lenses. Bigelow brusquely commented that if he wished to learn to make telescopes, he must go where they were made. Twenty years later, Bigelow was present when Alvan Clark received the Rumford Medal. Clark reminded him of their previous conversation, and said that he had indeed gone to where telescopes were made—to Cambridgeport![64]

Many people—sightseers as well as customers—visited the Clark factory and were shown every detail of the process; since the factory no longer exists and the Clarks left no records, all of our knowledge of the Clarks' methods comes from visitors' reports. One visitor to the Clark workshop thought the appliances both few and rude compared with those used by European artisans; he attributed the Clarks' success to skillful manipulation and personal supervision rather than reliance on precise mechanisms.[65] Another visitor was shown an old hen setting in a corner as an example of a manufactory quite as wonderful as the tele-

---

[61] "Remarks by Alvan Clark on Receipt of Rumford Medal," Proc., *American Academy of Arts and Sciences*, vol. 7 (1865–1868), pp. 244–249.

[62] Obituary of Alvan Clark, *Boston Post*, 20 August 1887, p. 8.

[63] Alvan Clark autobiography, op. cit., p. 116.

[64] Ibid.

[65] Ralph Copeland, "Notes on a Recent Visit to Some North American Observatories," *Copernicus*, vol. 3 (1884), p. 138.

scope workshop.[66] Maria Mitchell, to whom lens grinding seemed a tedious operation, told her students that "into every superior work the martyrdom must come." [67]

There was nothing unusual, unless perhaps their great carefulness, in the Clark technique for rough grinding and polishing a lens.[68] The blank disc was first polished so it could be tested for purity and evenness— a piece of glass too heavily striated would be rejected. The grinding and preliminary polishing were done on a rudimentary machine which consisted simply of a horizontal turntable rotated by steam power. The table held the tool, a cast iron lap of the same curvature as the lens, but reversed. The lens was held on the rotating lap and slowly moved about. Small lenses could be worked by one man; a larger lens was fitted with four wooden handles by which two workmen, walking around the table, could give the lens the proper movement. Alternately, the lens was supported from a horizontal beam which mechanically imparted a reciprocating motion.[69] The early Clark lenses were ground with emery, but by 1887 the Clarks were using cast iron sand as an abrasive because it had a lesser tendency to break down.[70] When the rough grinding was finished the metal lap was exchanged for one of pitch, and the lens was polished. The pitch was grooved to prevent it from sticking, and fed with water and rouge. In a letter written in 1892 the Clarks noted that a certain lens had indeed been polished with pitch and that they had "never used cloth polishers." [71] Small lenses were polished in the same way they were ground. Larger lenses were placed on the original metal lap but separated from it by a piece of Brussels carpet, and the pitch lap, which rested on the rotating lens, was moved back and forth.

Then came the delicate process of perfecting the lens by local correction—a method the Rumford Committee found both important and

---

[66] Garth Galbraith, "The American Telescope Makers," op. cit.

[67] Maria Mitchell, Alvan Clark and Telescope Making, loc. cit.

[68] The fullest discussion of the Clarks' methods of figuring a lens is in "The Alvan Clark Establishment," *Scientific American*, vol. 57 (1887), pp. 198–199.

[69] John F. Sullivan, "A Visit to Alvan Clark, Jr.," *Popular Astronomy*, vol. 35 (1927), p. 389.

[70] *Boston Journal of Chemistry*, vol. 6 (August 1871), p. 16. See also "The Alvan Clark Establishment," op. cit., p. 198.

[71] Alvan Clark & Sons to Elihu Thompson (sic), 25 November 1892 (letter in Thomson papers, American Philosophical Society).

original.[72] The Clarks had been using local correction for several years when Foucault in Paris announced his invention of this method in 1859. The idea for local correction might have been suggested to the Clarks by Henry Fitz of New York, from whom they had bought some pieces of optical glass.[73] Fitz retouched only one surface of his compound lenses, however, while the Clarks regularly retouched all four surfaces of each achromatic lens combination.

Local correction was used to remove the errors of figure remaining after the rough grinding and polishing were done. Because even the best glass available during the 19th century was irregular—as their tests with a polariscope showed—the Clarks used local correction to obtain the sharpest possible focus rather than mathematically true curves.[74] This method of figuring, therefore, lessened their dependence on absolutely homogeneous glass discs.

To locate the figure errors the Clarks developed a test similar and prior to Foucault's knife-edge test for mirrors. Their test was made either on an actual star or, more conveniently and more frequently, on an artificial star in the horizontal tunnel which stretched out 230 feet from the cellar of their workshop. The image of a point source of light was examined at the focus of the lens: a perfectly figured lens would appear uniformly illuminated while an imperfect lens would not.[75] The Clarks later devised and used a test twice as sensitive as this original one. Light from a point source was focused by the lens, and then reflected by an optically plane mirror back through the lens to an eyepiece close to the light source. Since light passed twice through the lens, the effect of any irregularities was doubled.

Tests of photographic lenses had to be made photographically. One method used by the Clarks involved photographing a star several times, with the plate at both sides of, and at varying distances from, the focus: a perfect lens would form evenly illuminated images. The Clarks also tested their photographic lenses by taking a spectrograph: a correctly

---

[72] "Remarks by Alvan Clark on Receipt of Rumford Medal," op. cit., p. 245.

[73] Henry Fitz's Account Book, 1851–1855 (in Smithsonian Institution, Division of Physical Sciences, Cat. 317,026).

[74] "Alvan Graham Clark," *Proc., American Academy of Arts and Sciences*, vol. 33 (1897–1898), p. 522.

[75] Simon Newcomb, "New Refracting Telescope of the National Observatory, Washington, D.C.," *Science Record*, (1874), p. 331.

figured lens imaged a spectrum as an even line, while an imperfect lens caused the width of the spectrum to vary.[76]

Once found, the location of the irregularities would be marked with a red powder. The lens was then laid flat and either Alvan Clark or Alvan Graham would retouch the offending areas. Each lens was, of course, tested and refigured innumerable times. Alvan Clark's sense of touch was said to be so acute that when a lens appeared perfect to the eye his fingers could still detect slight irregularities. For the final rub, Alvan Clark could find no sufficiently soft cloth, and so he used his bare thumbs! This also insured immediate detection of any particles of grit which might have gotten into the fine polishing powder. George Davidson noted that, when he last saw Alvan Clark, in 1885, the optician's thumbs had actually burst open from this punishing technique.[77]

Most of the Clark objectives were on the pattern of Fraunhofer lenses and were figured as simply as possible. The outer, crown glass lens was equiconvex, while the flint lens was biconcave, with the side toward the eyepiece nearly flat. The inner surfaces of the early objectives were ground with equal, but opposite, curves. Since this arrangement was found to produce an objectionable "object-glass ghost," in later instruments these curves were given slightly different radii. The inner surfaces of the two lenses were separated by a distance, depending on the aperture, which might be as much as several inches. With vents in the side of the tube, this separation allowed a free circulation of air between the components, so as to equalize more quickly the temperatures of the glass and the external air. Furthermore, it permitted easy cleaning of the lenses. This last convenience was in Alvan Clark's mind as early as 1851, when he patented an "Improvement in Telescopes" which consisted of "a simple and substantial eye-piece wherein ready access may be easily had to the glasses or lenses in order either to cleanse or repair them, as the case may require." [78] (See fig. 18, p. 61.)

The Clarks did not join in the contemporary mathematical search for more perfect lens configurations. In fact, there is little indication that they concerned themselves at all with the new forms described by theoreticians

---

[76] "The Alvan Clark Establishment," op. cit., p. 199.

[77] Obituary of Alvan Clark, *San Francisco Daily Examiner*, 20 August 1887, p. 1; this includes notes by Edward S. Holden and Ferdinand Clark. The latter's notes are quoted extensively in *English Mechanic*, vol. 46 (1887–1888), p. 83.

[78] U.S. Patent 8509 (11 November 1851).

such as Gauss, J. Herschel, Hastings, or Taylor. They made only one large lens—the 9-inch for Princeton—according to the Gaussian curves. They found these components, both of which are meniscus, more difficult to make, and disagreed that they gave a more complete achromatism and better definition than the more conventional ones.[79]

The Clarks, who almost always corrected their object glasses for the visual rays, had two methods of adapting these instruments for photography. The Lick telescope is provided with a third lens which can be added to the visual double achromat. The third lens was, of course, extremely expensive. In 1887, therefore, with the help of Edward C. Pickering, then director of the Harvard Observatory, the Clarks devised a new combination of two lenses which could be used for both photographic and visual observations. In this design the crown glass component is more convex on one side than the other. For visual work the flatter side is in contact with the flint lens; for photography the crown lens is reversed and moved farther forward.[80]

The Clarks were lens grinders par excellence, perhaps the most skillful the world has ever known. Quite understandably, therefore, Alvan Graham erred completely when he forecast the direction of 20th-century telescope developments. At the first meeting of the Congress of Astronomy and Astrophysics, held in Chicago during the Columbian Exposition of 1893, he read a paper on "Great Telescopes of the Future" in which he emphatically pronounced himself in favor of large refractors.[81] Glassmakers, he was sure, could cast optical discs greater than 40-inches in diameter. Using local correction, an optician—an artist in light and shade—could not only figure the large lenses but could also compensate for slight imperfections in the glass. The extent of the absorption of light through large lenses had been greatly exaggerated, as a recent experiment in his manufactory had shown. Finally, deformation of lenses under their own weight always occurred in a nearly compensatory manner. Alvan Graham showed a lack of awareness of the advances in

---

[79] Letter from Alvan Clark & Sons of 10 March 1879, quoted in Thomas Nolan, *The Telescope* (New York, 1881), p. 60.

[80] "Harvard Observatory and the Henry Draper Memorial," *Scientific American*, vol. 57 (1887), p. 278.

[81] Alvan Graham Clark, "Great Telescopes of the Future," *Astronomy and Astrophysics*, vol. 12 (1893), pp. 673–678. See also Alvan Graham Clark, "Possibilities of the Telescope," *North American Review*, vol. 156 (1893), pp. 48–53.

reflector construction which had been made during the previous thirty years. He spoke, for instance, only of speculum metal mirrors, disregarding silvered-glass mirrors because the problems of preserving and replacing the silver film were so formidable; Henry Draper and John A. Brashear, men with whom he was well acquainted, had, however, shown this type of mirror to be practicable. And just as Alvan Graham was stressing the impossibility of mounting a mirror so as to prevent flexures detrimental to good definition, George Willis Ritchey was solving that problem.

Until about 1880 the Clarks usually supplied the tubes, mounts, and various accessories, as well as the optical parts of telescopes; after that date small mounts for their lenses were frequently made by Fauth and his successors of Washington, D.C.,[82] and larger mounts were made by Warner & Swasey of Cleveland, Ohio.[83] The reasons for this change probably included the rise of these mechanical companies and George Clark's declining health.[84] The Clark mounts were usually of the German equatorial type and were furnished—money permitting—with a clock drive and graduated circles.

The Clarks were among the earliest telescope makers to use light metal for telescope tubes. A good tube should be strong, rigid, and lightweight. During the first half of the 19th century the common materials for tubes were wood and brass: wood was used, for instance, by Henry Fitz in America, and by Merz und Mahler for the Pulkowa and Harvard equatorials; brass was used for spyglasses, and by the Repsolds for the Oxford heliometer. Following suit, the Clarks used wood for such instruments as the $18\frac{1}{2}$-inch Dearborn and the 9-inch Yale refractors. As early as 1853, however, the Clarks were making some tubes of thin tin—or zinc for objectives of about 7 inches or more in diameter—overlaid with paper and paste until of a suitable thickness and until "tolerably straight," and then sanded. Finally, according to Alvan Clark, they gave the tube several coats of paint "of different tints, blue, red and yellow, all faint, but of the same depth when dry; they are rubbed down together. The effect is very pleasing to the eye, and I think, under copal varnish, this

---

[82] *Sidereal Messenger*, vol. 8 (1889), p. 192.

[83] The first Warner & Swasey telescope mount was for the $9\frac{1}{2}$-inch aperture refractor for Beloit College. See below, catalog of Clark instruments.

[84] "George Bassett Clark," *Proc., American Academy of Arts and Sciences*, op. cit., p. 362.

tu e must e very secure.

example of the 25-inch refracting telescope Thomas Cooke had made for R. S. Newall in England, and in 1871 they began making tubes of steel plates riveted together. These metal tubes weighed less, yet were stiffer than wooden ones.[86]

The Clarks—particularly Alvan and Alvan Graham—were known as practical astronomers as well as astronomical instrument makers. Besides drawing the above-mentioned map of the Orion nebula, they searched for double stars, participated in solar eclipse expeditions, made photometrical comparisons of the brighter celestial objects, and measured the lines in the auroral spectrum. These observations, although sporadic, were respected and often quoted in contemporary literature.

Before the development of the Hartmann test, the standard way of describing the perfection of a lens was in terms of its actual defining power. The Clarks tested each completed lens by searching for new and difficult double stars as well as by separating known binaries. Alvan Clark began by sweeping the southern skies—where he found some curious binaries—because he naively assumed that the great Struves, with the 9.6-inch Dorpat refractor, and later with the 15-inch at Pulkowa, must have discovered every double star in the northern hemisphere that was visible with smaller telescopes. When he eventually turned his telescope northward, Clark found several binaries the Struves had missed.[87] Without the benefit of astronomical background or library, Alvan Clark was ignorant of the novelty of some of his early observations. He wrote repeatedly to the Bonds, seeking confirmation of his discoveries; although helpful, the Bonds were hardly enthusiastic.

On 31 March 1851 the *Boston Daily Evening Transcript* published a letter from Alvan Clark reporting his recent discovery of an eighth-magnitude double star in Canis Minor, not listed in Smyth's *Celestial Cycle*. According to Clark, "This discovery (to which is attached a strong negative evidence of originality,) was the result of a systematic and laborious search, for unknown double stars: a search hitherto only

---

[85] Alvan Clark to William R. Dawes, quoted in Frederick Brodie, "Notes on the Manufacture of Tubes for Refracting Telescopes," *Monthly Notices, Royal Astronomical Society*, vol. 17 (1856–1857), pp. 35–36.

[86] *Boston Journal of Chemistry*, vol. 6 (August 1871), p. 16.

[87] William R. Dawes, "New Double Stars Discovered by Mr. Alvan Clark, Boston, U.S.; with Appended Remarks," *Monthly Notices, Royal Astronomical Society*, vol. 17 (1857), pp. 257–259.

rewarded, by learning with what skill and patience, observers have sought out, and recorded, the places and characters, of these numerous and interesting telescopic objects."[88] In search of more information about these stars, Clark sent a reprint to William Mitchell on Nantucket.[89] The Boston newspaper of 2 April carried another letter in which Alvan Clark explains that the double star had already been cataloged by Struve, "which adds anew to my dearly bought impression, that wary manage ment is necessary to win reprisals from the 'upper deep.' "[90] Later that month Clark informed the newspaper editor of a close approach of Jupiter and its satellites and the double star θ Virginis, which makes a "peculiarly beautiful group for the telescope."[91]

Later in 1851, in search of a more sympathetic and helpful audience than he could apparently find at home, Clark wrote to William R. Dawes in England. Dawes, who immediately appreciated Clark's work, published in the *Monthly Notices* measures of his binary stars, descriptions of his instruments, and the comment that Clark's "eye, as well as his tele scope must possess extraordinary power of definition."[92]

There is evidence that Clark's early observations were not unappreciated in his own country. Benjamin A. Gould and Joseph Winlock, both of whom had received as good an astronomical education and employment as was then possible, erected the temporary Cloverden Observatory in Cambridge in the early 1850's. In a prominently published description of their work, in 1854, they singled out the "experience of Mr. Alvan Clark" which supplied them with the positions of difficult double stars, many of which Clark had actually discovered.[93]

---

[88] Alvan Clark, "Search for New Double Stars," *Boston Daily Evening Transcript*, 31 March 1851, p. 2.

[89] Alvan Clark to William Mitchell, 1 April 1851 (letter in library of the Maria Mitchell Association).

[90] Alvan Clark letter to the editor, *Boston Daily Evening Transcript*, 2 April 1851, p. 2.

[91] Alvan Clark letter to the editor, *Boston Daily Evening Transcript*, 25 April 1851, p. 2.

[92] William R. Dawes, "New Double Stars Discovered by Mr. Alvan Clark . . .," op. cit. See also below, description of Dawes' apparatus in catalog of Clark instruments.

[93] Benjamin A. Gould and Joseph Winlock, "Cloverden Observatory and the Shelby Equatorial," *Proc.. American Association for the Advancement of Science*, vol. 8 (1854), p. 87.

Alvan Graham made similar observations. His most noted find, a classic case of serendipity, was the small companion of Sirius. The existence and position of this star had been predicted from the proper motion of Sirius, and the star had been frequently sought by others. Alvan Graham, however, was probably unaware of these previous researches.[94] On the evening of 31 January 1862 the Clarks—Alvan watching the time, and Alvan Graham at the ocular of the 18½-inch telescope—were trying to ascertain how long the light of Sirius was perceptible before the star itself was in view.[95] While Sirius was still behind the corner of a building Alvan Graham noticed the Pup, before it had been in the field for three seconds.[96] For this discovery he was awarded the 1862 Lalande Prize of the French Académie des Sciences. It is curious to note that although the Clarks designed and built micrometers, they seem not to have measured the star systems they discovered. As Dawes had done for Alvan, senior, S. W. Burnham, in 1879, published measurements of fourteen binaries discovered by Alvan Clark, junior.[97]

Extensive preparations were made to observe each solar eclipse which occurred during the second half of the 19th century. Besides making many of the instruments used to study the sun, the Clarks participated in several of these expeditions. George and Alvan Graham observed the total eclipse of 7 August 1869 at Shelbyville, Kentucky, as members of the United States Coast Survey party organized by Joseph Winlock of Harvard. Using the 5½-inch equatorial refractor the Clarks had recently made for the Harvard Observatory, George Clark and John A. Whipple, a Boston photographer, obtained eighty good pictures of the eclipse, three of which were taken during totality.[98] A chronograph, electrically connected with the shutter, recorded the time of each exposure. Alvan Graham's observations with a small refractor were undistinguished except for his confirmation of a flock of dark objects—presumably meteors—passing in front of the moon.[99]

---

[94] Simon Newcomb, "New Refracting Telescope of the National Observatory, Washington, D.C.," op. cit., p. 327.
[95] John Fulton, *Memoirs of Frederick A. P. Barnard* (New York, 1896), pp. 245–246.
[96] "Remarks by Alvan Clark on Receipt of Rumford Medal," op. cit.. p. 248.
[97] S. W. Burnham, "Double Stars Discovered by Mr. Alvan G. Clark," *American Journal of Science*, vol. 17 (1879), pp. 283–289.
[98] *U.S. Coast Survey Report, 1869*, p. 138.
[99] Ibid., pp. 136, 139.

Alvan Graham observed the eclipse of 22 December 1870 at Jerez de la Frontera, Spain, again as a member of a Coast Survey party directed by Winlock. His assignment was to help Winlock make spectroscopic observations of the solar corona. A spectroscope was attached to the above-mentioned 5½-inch Harvard equatorial. Winlock watched the spectroscope, while, in his words, "Mr. Alvan G. Clark, whose skill in everything pertaining to a telescope insured careful and judicious management of the instrument, was stationed at the finder to direct the telescope to the parts of the corona which were to be examined, and at the same time to observe incidentally general phenomena."[100] Their results were frustrated, however, by clouds.

On 29 July 1878 Alvan Graham was at Creston, Wyoming, a member of the United States Naval Observatory eclipse party organized by William Harkness. In cooperation with A. N. Skinner, he obtained six photographs of the corona, which were described as "at least as extensive and as rich in detail as any ever taken."[101] They used a camera with a 6-inch aperture Dallmeyer lens, supported on an equatorial mount pirated from an instrument made by the Clarks for the 1874 Transit of Venus.[102] Taking these pictures was really a two-man operation. Clark guided the camera and inserted and removed the sensitized plates; Skinner timed the exposure with a sand glass and regulated the exposures via a black wood fan with which he could cover the object lens.[103] George Clark declined invitations to join expeditions to observe these latter two eclipses.[104]

The three Clarks working together made photometrical experiments in 1862–63 in a unique way.[105] Their results for the sun and moon were not far from accepted modern values, but their stellar results were less reliable. Where others used a wedge or polar

---

[100] *U.S. Coast Survey Report, 1870*, p. 140.

[101] Edward S. Holden, "Astronomy," *Annual Record of Science and Industry* (1878), p. 21.

[102] See below, description of Transit of Venus apparatus, in catalog of Clark instruments.

[103] *U.S. Naval Observatory Reports of the Total Solar Eclipses of July 29, 1878, and January 11, 1880*, p. 51.

[104] John A. Brashear, "George Bassett Clark," *Astronomy and Astrophysics*, vol. 12 (1893), pp. 367–372.

[105] Alvan Clark, "The Sun a Small Star," *Memoirs, American Academy of Arts and Sciences*, vol. 8 (1863), pp. 569–572. See also Alvan Clark, "The Sun and Stars Photometrically Compared," *American Journal of Science*, vol. 36 (1863), pp. 76–82.

izing photometer to reduce the apparent brightness of the objects under consideration, the Clarks used a lens of very short focal length to enlarge the image. This lens was placed below the vertical shaft at the open end of their long underground testing chamber, and received sunlight from mirrors and totally-reflecting prisms. The solar image formed by the lens was viewed through a small hole in a brass strip, at such a distance that it appeared as bright as did another object at a specified distance. The Clarks explained that the sun, removed to ten times its present distance, would appear as bright as the solar image magnified ten diameters, and that either of these maneuvers would cause a hundred-fold decrease in brightness. They expressed their results for different objects in terms of the number of diameter reductions needed for comparable brightness. In other words they found, for instance, that the midday sun is 23.89 magnitudes brighter than Sirius and 13.01 magnitudes brighter than the moon; they saw only 1.11 magnitudes of difference between Sirius and Procyon and only 0.14 magnitudes difference between Castor and Pollux. Edward C. Pickering later took account of this work when he defined an average magnitude for the sun.[106]

The Clarks' only known recorded spectroscopic observation was made by Alvan Graham during a period of intensive auroral activity. On 24 October 1870, using a single-prism chemical spectroscope, he observed the spectrum of the aurora borealis and marked the position of four bright lines. Edward C. Pickering reduced his measurements to wave lengths and reported them to the editors of *Nature*.[107] For several years, while astronomers were seeking a consensus regarding the positions of auroral lines, Alvan Graham's measurements were repeatedly cited; his measures, however, differ considerably from the modern consensus.

As they got older, and as their business expanded, the Clarks hired other men to do at least the preliminary work on the instruments. The most valuable of their assistants was Carl Axel Robert Lundin,[108] a Swedish optician and mechanician who joined the Clarks as their chief instrument maker in 1874 at the age of twenty-three. He remained with them—

---

[106] Edward C. Pickering, "Dimensions of the Fixed Stars," *Proc., American Academy of Arts and Sciences*, vol. 16 (1880–1881), p. 2.

[107] Edward C. Pickering, "The Spectrum of the Aurora," *Nature*, vol. 3 (1870), pp. 104–105.

[108] "Carl Axel Robert Lundin," *Dictionary of American Biography* (New York, 1933), vol. 11, pp. 505–506.

and with their successors, the Alvan Clark & Sons Corporation—until his death 41 years later. The Clarks quickly recognized his talents, taught him their methods, and increasingly relied on him to finish their contracts. It is, again, difficult to know the extent of Lundin's work, especially as he seems to have been of an "exceedingly retiring and modest disposition."[109] After Alvan Graham's death in 1897 Lundin took over complete responsibility for the instrument construction, and advertisements for the corporation singled out Lundin as the optical expert. In 1905, upon completion of an 18-inch objective for Amherst College, Lundin received an honorary Master of Arts degree from Amherst, as had Alvan Clark half a century earlier. Other honors included a medal from the 1876 Centennial Exhibition, a diploma from the 1893 Columbian Exposition, fellowship in the American Association for the Advancement of Science, charter membership in the American Astronomical Society, and membership in the Swedish Society at Harvard.

The number of other workers in the Clark factory varied with the jobs at hand. In the fall of 1873 Simon Newcomb asked the Clarks not to diminish their work force after finishing the great Washington equatorial, as they would be needed to complete the Transit of Venus apparatus on time.[110] Soon thereafter the Clarks were employing a half-dozen hands and, even in the busiest times, were said not to have more than ten workers besides themselves.[111] Around 1881 the Clarks complained that "all the instrument makers are very busy and it is impossible to get all the workmen we want."[112] A few years later a visitor to their factory mentioned seeing four or five French opticians.[113]

Alvan Clark, Sr., was active in the business until just a few years before his death in 1887, at the age of eighty-three. Exactly how much work he had been doing in the later years is unclear. One account tells that his sons, after figuring and polishing a lens as best they could, brought it to him for the final examination and correction.[114] In 1884

---

[109] W. L. Watts to David Todd, 2 June 1905 (letter in Todd papers, Yale University Archives).

[110] Simon Newcomb to Alvan Clark & Sons, 15 October 1873 (letter in U.S. Naval Observatory Papers, U.S. Archives).

[111] An occasional correspondent of the Tribune, "Two Giant Telescopes," op. cit.

[112] Alvan Clark & Sons to Edward S. Holden, 1 September 1881 (?) (letter in Lick Observatory Archives).

[113] John F. Sullivan, "A Visit to Alvan Clark, Jr.," op. cit., p. 389.

[114] *English Mechanic*, vol. 46 (1887-1888), p. 83.

Alvan Clark sent a small object-glass to Amherst College, which he claimed to have reworked by his own hands entirely.[115] The following year Edward S. Holden found him "wonderfully well preserved" and noted almost with surprise that he could no longer do optical work.[116] Although he probably had little to do with the actual figuring of the Lick objective, Alvan Clark was alert and vigorous enough to participate in the testing of it in the summer of 1886! Charles A. Young reported that one time when he was in Cambridgeport Clark not only remained present during the whole evening, but also "examined every object, even when the position required was such as to cramp and try younger limbs and muscles. His eye seems to have lost little of its original keenness, though trembling hands and easy weariness no longer permit him to do much actual labor."[117] Around 1885 the "venerable old man" returned to his paint brushes, which he had not touched for twenty-five years; he painted at least three large oil portraits—of his two sons, and of his only grandson, Alvan Clark III.[118]

George was much less robust than his father and survived him by only four years. As early as 1887 an observer thought that Alvan Graham was the responsible and active member of the firm, "if his movements can be called by that title."[119] In the fall of 1891 the brothers dissolved their partnership because George felt physically unable to continue working.[120] Alvan Graham did most of the work on the Lick lens; after that he seems to have merely supervised the construction of Clark lenses while Lundin did the actual work. Alvan Graham's last project, in the spring of 1897, was to accompany the 40-inch lens to the Yerkes Observatory and superintend its mounting there.

After the death of Alvan Graham, the Alvan Clark & Sons Company continued, without any Clarks, for another sixty years. George Clark and his sisters were childless; Alvan Graham's only son died at the age

---

[115] Alvan Clark to David Todd, 7 October 1884 (letter in Todd Papers, Yale University Archives).

[116] Edward S. Holden to Richard S. Floyd, 23 June 1885 (letter in Lick Observatory Archives).

[117] Charles A. Young, "A Look Through the Great Object-Glass," *The Popuar-Science News*, vol. 20 (1886), p. 138.

[118] *Science*, vol. 7 (1886), pp. 303-304.

[119] Charles Plum to the Lick Trust, 21 May 1887 (letter in Lick Observatory Archives).

[120] *Publications, Astronomical Society of the Pacific*, vol. 3 (1891), p. 377.

of fourteen, and his daughters did not marry instrument makers. While Lundin was active, the company made several large and notable telescopes; later they were called upon for smaller instruments for teaching and amateur purposes. In 1933 the Clark concern was taken over by the Sprague-Hathaway Mfg. Co., of West Somerville, Massachusetts.[121] Eight years later the Clark Company announced that the Perkin-Elmer Corporation was to act as consultants in connection with the design and construction of its optical systems.[122] This assistance, however, was not enough, and the demise of Sprague-Hathaway, in 1958, carried with it the dissolution of Alvan Clark & Sons.

---

[121] *Popular Astronomy*, vol. 42 (1934), p. 608.

[122] See advertisements for Alvan Clark and Sons Company in *Popular Astronomy*, vol. 49 (1941).

Part II

Catalog of Astronomical Instruments Made and Remade by Alvan Clark & Sons, 1844–1897

This catalog includes all Clark instruments which have come to my attention; omissions—there may be a great many—are unintentional. The purpose of the catalog is twofold. Primarily it shows the extent to which astronomical observatories built during the 19th century, especially in America, were equipped with at least some apparatus from the Clark workshop. The second purpose of the catalog is to tell more about the Clarks, for stories pertaining to the various instruments often reveal personal characteristics as well as the professional accomplishments of the makers.

The catalog is arranged alphabetically by name of instrument owners, both private persons and institutions; instruments which had several owners are cross-referenced. Instruments of all sizes, and accessories as well as major telescopes, are included in the catalog. In many cases the smaller and more obscure instruments, simply because of their obscurity, are described in greater detail than the better known ones.

MRS. EDWIN A. ABBEY owned a 5-inch aperture refracting telescope mounted on a tripod, the lens of which was reputed to have been made by Alvan Clark himself in 1871. Mrs. Abbey, nee Mary Gertrude Mead, was a Vassar graduate and undoubtedly a student of Maria Mitchell, as Miss Mitchell placed the order for the telescope. This instrument is now at the Maria Mitchell Observatory (q.v.).[1]

In 1873 ABBOT ACADEMY, a girls' school in Andover, Massachusetts,

---
[1] From private correspondence with Dorrit Hoffleit of the Maria Mitchell Association.

purchased a 7½-inch equatorial refractor from the Clarks for $1150. Owing to their limited budget they had to choose between graduated circles and a driving clock—either of which cost an additional $500.[2] Apparently, they found money for both.

In 1888 ALBION COLLEGE, in Albion, Michigan, built an astronomical observatory for the use of its students. The equipment included an 8-inch aperture Clark refractor which was equatorially mounted and provided with driving clock, divided circles, and filar micrometer.[3]

By 1866 the Clarks had supplied ALLEGHENY COLLEGE, in Meadville, Pennsylvania, with a 7-inch refracting telescope which, although equatorially mounted, was not equipped with either graduated circles or clock drive.[4]

ALLEGHENY OBSERVATORY of the University of Pittsburgh used the Clarks' services during the 1870's; thereafter they relied on the local and talented John A. Brashear. The 13-inch Fitz lens of the Allegheny equatorial was stolen and slightly damaged in 1872. After its recovery the Clarks refigured it, making it, in terms typically used to describe Clark lenses, "l'un des meilleurs objectifs de cette dimension que l'on connaisse." [5] They also made a 4-inch photographic correcting lens which was used in conjunction with the 13-inch. Samuel Pierpont Langley, then doing solar physics at Allegheny, was using spectroscopic apparatus too heavy to be easily moved; to convert the equatorial into a heliostat the Clarks made a 12-inch diameter silvered plane mirror.[6] Other Clark instruments at Allegheny included a 3-inch objective lens [7] and a rock salt prism identified as having been figured by George Clark.[8]

The AMERICAN ACADEMY OF ARTS AND SCIENCES in Boston owned

---

[2] Mary Delcher of Abbot Academy to Maria Mitchell, 2 April 1872 (letter in library of Maria Mitchell Association).

[3] *Year Book of Albion College for 1888–89*, p. 138.

[4] Elias Loomis, *Practical Astronomy* (New York, 1866), 7th ed., pp. 496–497.

[5] Charles André and A. Angot, *L'Astronomie pratique et les observatoires en Europe et en Amérique*. Part 3: *États-Unis d'Amérique* (Paris, 1877), p. 126.

[6] Edward S. Holden, "Astronomy," *Annual Record of Science and Industry* (1877), p. 38.

[7] Alvan Clark & Sons to Samuel P. Langley, 5 November 1900 (letter in Smithsonian Institution Archives).

[8] John A. Brashear, "A Practical Method of Working Rock Salt Surfaces for Optical Purposes," *Proceedings, American Association for the Advancement of Science*, vol. 34 (1885), p. 76.

FIGURE 12.—Cambridge, Massachusetts, showing the location of the Alvan Clark & Sons establishment along the Charles River (circle at lower right). Prospect Street (circle at center right), where the Clarks lived from 1836 to 1860, is also indicated. Courtesy Library of Congress.

several pieces of scientific apparatus and made them available to scholars for research purposes. One of these instruments was a Clark-made micrometer-level which incorporated a telescope of 4 centimeters aperture, supported in Y's and moved by a vertical micrometer screw.[9]

---

[9] "List of Apparatus Relating to Heat, Light, Electricity, Magnetism, and Sound, Available for Scientific Researches Involving Accurate Measurements," *Annual Report . . . Smithsonian Institution . . . 1878*, p. 430.

The AMERICAN ASSOCIATION OF VARIABLE STAR OBSERVERS is the present owner of the 4-inch Clark telescope originally owned by William Tyler Olcott (q.v.). According to Walter Scott Houston, the present user of the telescope, on a good night it will show stars of fourteenth magnitude.

AMHERST COLLEGE, in Massachusetts, installed a Clark telescope of 7¼ inches aperture in 1854.[10] This telescope, as David Todd noted, was probably the first complete equatorial built and sold by the Clarks.[11] Moreover, it was very likely the first telescope ever provided with a clock drive regulated by a Bond spring-governor. The graduated circles, 12 inches in diameter, were read by verniers—the right ascension to 2 seconds of time, and the declination to 30 seconds of arc. The total cost of the telescope was $1800. In appreciation for his work on this telescope, and for his discovery of two new double stars while testing it, Amherst gave Alvan Clark an honorary masters degree in 1854.

Thirty years later Alvan Clark refigured the objective of Amherst's old transit instrument and claimed that he had done the work "by my own hands entirely."[12] This must have been one of the last pieces of optical work for which he was responsible.

As mentioned above (p. 000), Carl A. R. Lundin of the Alvan Clark & Sons Corporation figured the 18-inch object-glass for Amherst and was given an honorary masters degree in 1905.

ANTIOCH COLLEGE, in Yellow Springs, Ohio, although without an observatory, in 1878 reported having a "very good" Clark refracting telescope of 4.94 inches aperture, equatorially mounted on a tripod.[13]

Several times during the latter half of the 19th century lists were compiled of American astronomers, observatories, and telescopes. Although most of the instruments listed can be verified elsewhere, a few cannot. In 1877 BATES COLLEGE, in Lewiston, Maine, was reputed to

---

[10] Elias Loomis, *Recent Progress of Astronomy* (New York, 1856), pp. 264, 391.

[11] David Todd, "Early History of Astronomy at Amherst College," *Popular Astronomy*, vol. 11 (1903), p. 324.

[12] Alvan Clark to David Todd, 7 October 1884 (letter in Todd Papers, Yale University Archives).

[13] Edward S. Holden, "Astronomy," *Annual Record of Science and Industry* (1878), p. 76.

have a Clark refractor of 6¼ inches aperture;[14] astronomers at Bates College today, however, have no knowledge of this instrument.[15] The reference is very likely to the telescope Oliver C. Wendell (q.v.) had recently acquired and which he had not yet installed in his private observatory.

R. R. BEARD of Pella, Iowa, was a banker with an amateur interest in astronomy. By 1887 he had acquired a Clark 6½-inch aperture equatorial refractor[16] which, in 1904, he donated to the local Central College (q.v.).

The first astronomical telescope built by Warner & Swasey was erected at BELOIT COLLEGE in Wisconsin in 1882. This equatorial mount was designed to hold a 9½-inch Clark objective.[17] In 1916 the Clark lens was remounted by Warner & Swasey; this telescope remained at Beloit until 1967, when it was sold to Dave Garroway (q.v.). The original Warner & Swasey mount is in the Smithsonian Institution (q.v.).

By 1888 F. G. BLINN of East Oakland, California, had a 5-inch Clark achromatic refractor, equatorially mounted, with circles, slow motion, and an effective battery of eyepieces.[18]

On BONNER'S HILL, just outside the city of Quebec, the provincial government erected an astronomical observatory in 1864, mainly for the purpose of accurate timekeeping.[19] The novel and convenient arrangement for a shutter over the slit of the dome was suggested to Commander E. D. Ashe, R.N., by Alvan Clark, who used canvas stretched between deal rods. These rods, which were fixed with hooks at the top, were springy enough to hold the canvas close against the dome. Ashe replaced the canvas with a thin deal board and was very pleased with the results. The original equipment of the observatory included an 8-inch Clark equatorial, of 9-feet focal length. After the demise of the Bonner's Hill observatory, in 1874, the Clark telescope was placed in the new observa-

---

[14] "Size of the Principal Telescopes in the World," *Popular Science Monthly*, vol. 10 (1876–1877), p. 576.

[15] From private correspondence with Karl S. Woodcock, Emeritus Professor of Physics and Astronomy at Bates College.

[16] *Sidereal Messenger*, vol. 6 (1887), p. 78.

[17] Ibid., vol. 1 (1882), p. 200.

[18] Edward S. Holden, *Handbook of the Lick Observatory* (San Francisco, 1888), p. 122.

[19] E. D. Ashe, "Description of the Observatory, Bonner's Hill, Quebec," *Monthly Notices, Royal Astronomical Society*, vol. 25 (1864–1865), p. 29.

FIGURE 13.—Alvan Clark & Sons factory. The telescope mount and tube in the foreground were used for testing objective lenses. From cover of *Scientific American*, 24 September 1887.

tory in the Battlefield Park. When that observatory was demolished, in 1935, the telescope was given to the Collège des Jésuites (q.v.) in Quebec, where it is still being used.[20]

Toward the end of the century the equipment of the BOSTON UNIVERSITY observatory included a 5-inch Clark refractor, equatorially mounted on a tripod.[21]

ROLAND B. BOURNE of West Hartford, Connecticut, is the present owner and user of the finder scope of the 9.4-inch Clark equatorial re-

---

[20] From private correspondence with Bernard Weilbrenner, Quebec Provincial Archivist.
[21] *Boston University Yearbook* (1905), p. 71.

fractor previously owned by D. W. Edgecomb and Charles P. Howard (qq.v.). According to Bourne this 2-inch aperture instrument resolves well within the Dawes limit, and its surfaces are accurate to about 1/20 of a wavelength.

The Clark lenses retained their value for many years, and a used lens was often as highly prized as a new one. In 1856 FREDERICK BRODIE, in England, replaced his 6⅓-inch Munich refractor with a Clark objective of 7½ inches aperture.[22] This was the first lens which Clark had sold to Dawes (q.v.); in 1873 it was installed in the private observatory of Wentworth Erck (q.v.).

ADDISON BROWN, amateur astronomer, used his 4-inch Clark refractor, mounted on a tripod, to view the total solar eclipse of 1878.[23]

One of the last large instruments made completely by the Clarks was the 10-inch equatorial at BUCKNELL COLLEGE in Lewisburg, Pennsylvania.[24] This instrument, installed in 1887, is still in use.[25] It was provided with 5 common eyepieces, a polarizing eyepiece for solar observations, a micrometer, and a 2½-inch finder. The right ascension and declination circles were read with small telescopes. The observer could regulate the driving clock—which was changed only a few years ago—at will and move the telescope with handles without leaving the eyepiece.

The most productive and famous small Clark telescope was the 6-inch refractor made for SHERBURNE WESLEY BURNHAM in 1870. With this instrument Burnham began his study of the heavens and discovered several hundred double stars, many of which had been missed by more practiced astronomers with much larger telescopes.[26] In 1881 this 6-inch was mounted in the Students' Observatory at the University of Wisconsin (q.v.)

JACOB CAMPBELL, a New York City banker, owned a 12-inch equatorial refractor made by Henry Fitz. As they did with so many of the

---

[22] Frederick Brodie, "Notes on the Manufacture of Tubes for Refracting Telescopes," *Monthly Notices, Royal Astronomical Society*, vol. 17 (1856–1857), p. 33.

[23] *U.S. Naval Observatory Reports of the Total Solar Eclipses of July 29, 1878, and January 11, 1880* (Washington, D.C., 1880), p. 142.

[24] W. C. Bartol, "The Observatory's Equipment," *The University Mirror* (May 1887), pp. 100–101.

[25] From private correspondence with Robert R. Gross, Bucknell University Archivist.

[26] Letter from S. W. Burnham, *English Mechanic*, vol. 13 (1871), pp. 488–489.

large Fitz lenses, the Clarks reworked Campbell's objective.[27] The telescope tube—which is wooden, and probably the original one from the Fitz shop—now carries a plaque at the eye-end reading "ALVAN CLARK & SONS CAMBRIDGEPORT, MASS. 1867." This instrument was later sold to S. V. White (q.v.), and is now at Wellesley College (q.v.).

CARLETON COLLEGE, in 1878, purchased an 8¼-inch Clark equatorial. As the young college in Northfield, Minnesota, did not then have the requisite $3000, the Clarks agreed to wait a year and a half for the money.[28] Eight years later the Clarks made a third, photographic correcting lens, of the same aperture, and remodeled the mount to correspond to the shortened photographic focal length. This triple lens combination gave excellent star images over a field of 40 minutes radius. Guiding was done with a 5-inch telescope, of the same focal length as the triplet, provided with the original Clark micrometer.[29] With an enlarging lens made by Brashear in 1892, this instrument yielded a fine series of solar photographs.[30] The Clarks also provided Carleton with a chronograph [31] and a 4.8-inch object-glass for their 1885 Repsold meridian circle.[32]

The Beard Observatory of CENTRAL COLLEGE in Pella, Iowa, is named after R. R. Beard (q.v.). By 1899 the college could advertise that its students were permitted access to Beard's private observatory and his 6½-inch Clark equatorial refractor; [33] and five years later they announced that Beard had donated his telescope to the college.[34]

Since 1922 the Morrison Observatory (q.v.) has been affiliated with CENTRAL COLLEGE in Fayette, Missouri.[35]

---

[27] Alvan Clark to Edward S. Holden, 8 March 1876 (letter in Lick Observatory Archives).

[28] Delavan Leonard, *The History of Carleton College* (Chicago, 1904), p. 202.

[29] *Publications, Goodsell Observatory of Carleton College*, vol. 5 (1917), p. v.

[30] Ibid., vol. 3 (1901), pp. 1-2.

[31] Ibid., vol. 2 (1901), p. 2.

[32] [William W. Payne] "The New Repsold Meridian Circle of Carleton College Observatory," *Sidereal Messenger*, vol. 6 (1887), p. 304.

[33] *Catalogue of the Central College, Incorporated* (Central University of Iowa) (1899), p. 10.

[34] *The Alumni Record* (Central University of Iowa), vol. 2 (1904), no. 5, p. 6.

[35] Robert R. Fleet, "Dedication of the Morrison Astronomical Observatory," *Popular Astronomy*, vol. 44 (1936), pp. 476-480.

The CENTRAL MISSOURI AMATEUR ASTRONOMERS, in Moberly, Missouri, are the present owners of Robert Green's (q.v.) 4-inch Clark refractor.

In 1883 ANTHONY CHABOT donated a well-equipped astronomical observatory to the Board of Education of Oakland, California. From Fauth & Co. Chabot obtained a transit instrument and a chronograph. From the Clarks he bought an equatorial refracting telescope of 8 inches clear aperture. This telescope had large setting circles graduated on silver and read by verniers and microscopes, the hour circle to 1 second of time, and the declination circle to 5 seconds of arc. The battery of eyepieces ranged from 40 to 800 power; the Clarks seldom supplied eyepieces which magnified more than 100 times for each inch of aperture. Among the accessories of the Chabot telescope were a position micrometer and a spectroscope provided with both a prism and a plane diffraction grating.[36] Although moved several times to avoid the glare of city lights, this telescope is still in use.

The original UNIVERSITY OF CHICAGO provided the first home for the 18½-inch Dearborn telescope (q.v.), while the present University of Chicago encompasses the Yerkes Observatory (q.v.).

The 11-inch CINCINNATI equatorial, which O. M. Mitchel had purchased in Munich 35 years earlier, was worked over by the Clarks in 1876. They refigured, and improved, the objective; replaced the driving clock; added wooden arcs, visible from the ocular, to facilitate finding a celestial object; supplied an achromatic and a diagonal eyepiece; and they "neatly painted" the whole telescope and stand.[37] About the same time, the University of Cincinnati Observatory acquired from the Clarks an equatorially mounted comet seeker of 4 inches aperture.[38] Their 1889 meridian circle incorporated a 5½-inch Clark objective and a Fauth mount.[39]

The CINCINNATI ASTRONOMICAL SOCIETY inherited an 8¼-inch aperture equatorial refractor from Andrew Henkel (q.v.). Except for an

---

[36] *Annual Report of the Public Schools of the City of Oakland, for the Year Ending June 30, 1891*, pp. 31–33.

[37] *Annual Report of the Directors of the University of Cincinnati* (1877), p. 12.

[38] *U.S. Naval Observatory Reports of the Total Solar Eclipses of July 29, 1878, and January 11, 1880* (Washington, D.C., 1880), p. 235.

[39] *Appleton's Annual Cyclopaedia*, vol. 14 (1889), p. 42.

47

electric drive to replace the original gravity-driven one, the telescope has not been altered since it was made by the Clarks in 1880.

In 1872 COLUMBIA COLLEGE, in New York City, built a small astronomical observatory for educational and scientific work. It was furnished with a 6-inch aperture equatorial refractor, moved by clockwork, and an astronomical spectroscope, both made by the Clarks. The spectroscope held 7 heavy flint glass prisms through which light passed twice.[40] Columbia has been credited with several other 19th-century Clark instruments as well. J. K. Rees, then director of the Columbia observatory, viewed the 1882 Transit of Venus through a Clark 5-inch equatorial refractor moved by clockwork and "similar in all respects to the instruments made for the Transit of Venus expeditions of 1874" (q.v.).[41] And, according to an inventory of large telescopes, published by the *Sidereal Messenger* in 1884, there was then an 11-inch Clark refractor at Columbia.[42]

The national Argentinian observatory at CÓRDOBA, founded by Benjamin A. Gould in 1870, had two pieces of Clark apparatus. One was an equatorial mount for the 11-inch photographic lens made by Lewis Morris Rutherfurd—and not by Henry Fitz as has been often stated—in 1864;[43] this mount was clock driven and provided with a position circle micrometer. The other Clark instrument at Córdoba was a 5-inch equatorial refractor equipped with finely divided circles, but without clockwork.[44]

CORNELL UNIVERSITY built a temporary astronomical observatory in 1888. Among the original, and now misplaced, instruments was a 4½-inch Clark equatorial refractor.[45]

Charles A. Young at DARTMOUTH COLLEGE relied on the Clarks to make the apparatus for his studies of the solar spectrum. He apparently

---

[40] *Catalogue of the Officers and Students of Columbia College, for the Year 1872–1873* (New York, 1872), pp. 107–108. See also, "Size of the Principal Telescopes in the World," *Popular Science Monthly*, vol. 10 (1876–1877), p. 576.

[41] J. K. Rees, "Observations of the Transit of Venus, December 6, 1882," *Annals, New York Academy of Sciences*, vol. 2 (1880–1883), pp. 384–390.

[42] [William W. Payne], "Large Telescopes of the World," *Sidereal Messenger*, vol. 3 (1884), p. 194.

[43] Lewis M. Rutherfurd, "Astronomical Photography," *American Journal of Science*, vol. 39 (1865), pp. 304–309.

[44] *Resultados del Observatorio Nacional Argentino in Córdoba*, vol. 2 (1881), pp. xiii, lxiii.

[45] *The [Cornell] Register* (1888–1889), p. 35.

spent a lot of time in Cambridgeport, conferring with the Clarks on the details of the instruments. After several of these visits Young wrote articles—often for the *Boston Journal of Chemistry*—describing the work in progress at the Clark factory.

Shortly after his appointment to the college faculty in 1866 Young ordered a $350 spectroscope from the Clarks.[46] Although frequently modified, this instrument was in use until 1963; it is now preserved in the Dartmouth College Museum. Young used this spectroscope during the total solar eclipse of 1869 for his observations of the recently found green coronal line, and his discovery of a spectroscopic method of determining the moment of contact between sun and moon. As his observations might not have been possible with an instrument of lower dispersion, Young carefully described his apparatus. The spectrum was formed by a train of five heavy flint glass prisms of 45° each; and the prism box could be adjusted for position of best definition for the various lines. For comparison, a spark spectrum could be thrown into the field of view. The collimator and eye telescope were of $2\frac{1}{4}$-inch aperture and $16\frac{1}{2}$-inch focus. The eyepiece was provided with a micrometer for determining the position of new lines. Finally, the whole spectral apparatus was attached to a small comet seeker and equatorially mounted.

Upon his return from the eclipse expedition Young designed, and asked the Clarks to make, another spectroscope.[47] George Clark did most of the work on this instrument and suggested several convenient adjustments. This spectroscope had the dispersive power of 13 prisms of heavy flint glass, yet was light enough to be attached to Dartmouth's 6.4-inch German equatorial. Light from the slit and collimator passed through the lower portion of 6 prisms and one half prism; it was then reflected by a 90° prism back through the upper portion of the prism-train to the observing telescope, which was above and parallel to the collimator. One of George Clark's contributions was a new method of adjusting the

---

[46] Charles A. Young, "On a New Method of Observing Contacts at the Sun's Limb, and Other Spectroscopic Observations During the Recent Eclipse," *American Journal of Science*, vol. 48 (1869), pp. 370–378. See also the same author's "On a New Method of Observing the First Contact of the Moon with the Sun's Limb at a Solar Eclipse by Means of the Spectroscope" and "Spectrum Observations at Burlington, Iowa, during the Eclipse of August 7, 1869," *Proc., American Association for the Advancement of Science*, vol. 18 (1869), pp. 82–84, 78–82.

[47] Charles A. Young, "A New Form of Spectroscope," *Journal, Franklin Institute*, vol. 60 (1870), pp. 331–334.

prisms to their angle of minimum deviation, which depended upon altering the circumference around which the prisms stood. If not theoretically exact, this method, according to Young, was simple and solid and worked well. Young took this spectroscope to Spain for the 1870 solar eclipse, where he discovered the reversing layer of the sun. After the eclipse Young had the Clarks remodel this spectroscope to incorporate further improvements. He then thought this spectroscope inferior to very few then in use, and as optically perfect as any he had ever seen.[48] The spectroscope has since been retired and is on exhibit in the Dartmouth College observatory. (See fig. 19, p. 64.)

Dartmouth College, during Young's tenure, also acquired a Clark equatorial refractor. The 9.4-inch objective was ground according to Littrow's curves—i.e., a nearly equi-convex crown lens and a nearly plano-concave flint. According to Young, the spherical aberration was very perfectly corrected; the chromatic aberration was a little undercorrected, a pleasant contrast to the overcorrected Munich lens he had been using.[49] The tube was one of the earliest composed of plates of steel riveted together. The telescope was remounted and provided with a photographic correcting lens by the Alvan Clark & Sons Corp. around 1908.[50] (See fig. 20, p. 89.)

The first astronomical observatory in California was erected by GEORGE DAVIDSON in 1879 in LaFayette Park, San Francisco.[51] Its main instrument was an equatorial refractor, with a 6½-inch Clark object-glass, and a Fauth mount. Fauth & Co. had exhibited this instrument at the 1876 Centennial Exhibition in Philadelphia, where it won a gold medal. This telescope has recently been given to the California Academy of Sciences, of which Davidson was president for 14 years.

WILLIAM RUTTER DAWES, the Clarks' first important customer, was an innovative observer who would rather order new instruments than modify old ones. According to Simon Newcomb, "Mr. Clark, not being

---

[48] Charles A. Young, "The So-called Elements," *Knowledge*, vol. 1 (1881), pp. 151–152.

[49] Charles A. Young, "The Color Correction of Certain Achromatic Object-Glasses," *American Journal of Science*, vol. 29 (1880), pp. 454–456.

[50] John K. Lord, *A History of Dartmouth College, 1815–1909* (Concord, N.H., 1913), pp. 609–610.

[51] "George Davidson," *Dictionary of American Biography* (New York, 1930), vol. 5, p. 92.

FIGURE. 14.—William Rutter Dawes (1799–1868), correspondent, client, and friend of Alvan Clark. Courtesy Wellesley College Observatory.

a trained engineer, Mr. Dawes found in him one who was ready to adopt and incorporate in an instrument any feature he might desire, and who would follow his multiplicity of minute directions with the most scrupulous accuracy." [52] All in all Dawes purchased and used five different Clark instruments.

The first is well described in Alvan Clark's own words: "In 1853 I had finished a glass of seven and one half inches aperture, with which

---

[52] Simon Newcomb, "The Story of a Telescope," *Scribner's Monthly*, vol. 7 (1873–1874), p. 46.

the companion of 95 Ceti was discovered. Upon reporting this to Mr. Dawes, he expressed a wish to possess the glass, but to test its qualities further sent me a list of Struve's difficult double stars, wishing me to examine them which I did and furnished him with such a description of them as satisfied him that they were well seen." [53] Clark *reluctantly* sold this telescope to Dawes in March 1854 for $950.[54] Dawes used it for two years, mainly for observations of Saturn, and publicly declared it decidedly superior in illuminating and separating power to his old favorite, a Munich $6\frac{1}{3}$-inch refractor.[55] William Lassell was "astonished" by the quality of this telescope.[56] This lens was later used by Frederick Brodie (q.v.) and Wentworth Erck (q.v.).

The tube Clark made for this telescope was novel and successful enough to merit a description in the pages of the *Monthly Notices*. It was constructed of a sheet of zinc covered with brown paper; and although only $\frac{2}{10}$-inch thick, the tube was "remarkably solid and hard to the touch."[57] When he sold the $7\frac{1}{2}$-inch lens to Brodie, Dawes kept the tube and used it to carry another Clark lens.

In October 1855 Clark sent Dawes a second telescope—of 8 inches aperture and 10 feet focal length—which Dawes found as praiseworthy as the first.[58] Dawes used it for a few years but found it inconveniently long for his dome. It did its finest work in the hands of William Huggins (q.v.), to whom it was sold in 1858 for £200, the same price it had cost Dawes originally.

Dawes asked Clark to replace this lens with one of the same aperture, but shorter focus, and make a complete equatorial mount with clock drive. The new lens was almost completed when, during testing, Clark dropped and broke it. Nothing daunted, Clark immediately ordered another pair of discs from England; before they arrived, however, he had

---

[53] Autobiography of Alvan Clark published, among other places, in *Sidereal Messenger*, vol. 8 (1889), p. 114.

[54] Elias Loomis, *Recent Progress of Astronomy* (New York, 1856), pp. 390–391.

[55] William R. Dawes, "On the Telescopic Appearances of Saturn With a $7\frac{1}{2}$-inch Object-Glass," *Monthly Notices, Royal Astronomical Society*, vol. 15 (1854–1855), pp. 79–80.

[56] Elias Loomis, *Recent Progress of Astronomy*, p. 391.

[57] Frederick Brodie, "Notes on the Manufacture of Tubes for Refracting Telescopes," *Monthly Notices, Royal Astronomical Society*, vol. 17 (1856–1857), p. 35.

[58] William R. Dawes, "Telescopic Appearances of the Planet Saturn," *Memoirs, Royal Astronomical Society*, vol. 26 (1858), pp. 9–18.

found a suitable pair of optical discs in New York. He figured both pairs, and with them he discovered eight new double star systems.[59] He brought both finished objectives and one equatorial mount with him when he visited Dawes in 1859.[60] Both objectives were so fine that Dawes was puzzled to decide which to keep.[61] He finally sold the 8-inch aperture American glass, with the equatorial mount, for £500, and the Clarks sent him another mount for the 8¼-inch lens. The 8¼-inch was sold, in 1864, to Rev. R. E. Lowe (q.v.), who, seven years later, sold it to J. M. Wilson of Rugby School (q.v.). The definition of this object glass was declared, by Otto Struve, to be superior to that of the Dorpat and Pulkowa instruments.[62]

Heretofore Dawes had known the Clarks only as opticians and astronomical observers. His decision to replace his Munich equatorial mount with one made by the Clarks was prompted by the Bonds' enthusiastic report of the performance of the new clock drive they had made for the Harvard equatorial (q.v.).[63]

The Clarks' cast iron mount was so firm, compact, and elegant Dawes thought fit to describe it to the Royal Astronomical Society.[64] The tube was supported by the open ends of a movable arc structure which served as the polar axis. The intervening semicircular space held most of the wheelwork of the driving clock. Since the arc slid in a groove in the lower part of the mount, the polar axis could be easily adjusted for any latitude. With this mount, as opposed to one of the German form, stars could be followed continuously as they crossed the meridian. The motion of the driving clock, nicely regulated by a half seconds pendulum and a spring governor, was so equable that, even with high powers, Dawes

---

[59] William R. Dawes, "New Double Stars Discovered by Mr. Alvan Clark," *Monthly Notices, Royal Astronomical Society*, vol. 20 (1859), p. 55.

[60] Simon Newcomb, "The Story of a Telescope," *Scribner's Monthly*, vol. 7 (1873–1874), p. 46.

[61] William R. Dawes to George Knott, 15 August 1859 (letter published in *Observatory*, vol. 33 [1910], pp. 352–353).

[62] William R. Dawes to R. E. Lowe, 20 June 1864 (letter in Rugby School Archives).

[63] William R. Dawes to William C. Bond, 30 June 1857 (letter in Bond Papers, Harvard University Archives).

[64] William R. Dawes, "Description of an Equatorial Recently Erected at Hopefield Observatory, Haddenham, Bucks," *Monthly Notices, Royal Astronomical Society*, vol. 20 (1859–1860), pp. 60–62.

could keep a star under a micrometer wire for long periods of time. This telescope was equipped with a double eyepiece micrometer and "an ingenious apparatus for sketching the solar spots somewhat on the principle of a Camera Lucida." (See fig. 21, p. 91.)

In 1857 Dawes sold his $6\frac{1}{3}$-inch objective and tube, which had been made by Merz of Munich, and bought from the Clarks a new telescope to fit the German equatorial mount. With this 7-inch instrument Dawes and Lassell one evening were able to see Enceladus, and Lassell expressed surprise that this Saturnian satellite was so readily apparent in such a small telescope. When Alvan Clark arrived in England with the new instruments in 1859, Dawes sold the old telescope to George Knott (q.v.).[65]

The $18\frac{1}{2}$-inch DEARBORN equatorial, the Clarks' first world's largest, had a stormy history. It was ordered by the University of Mississippi (q.v.) which, under the presidency of F. A. P. Barnard, was strengthening its science facilities. An astronomical observatory—an exact replica of Pulkowa—was built in Oxford, Mississippi, in 1859.[66] With full confidence in the Clarks' ability, Barnard asked them to make a telescope of 19 inches aperture. He even offered to comply with the usual terms: $\frac{1}{3}$ down, $\frac{1}{3}$ when the objective was tested, and $\frac{1}{3}$ when the instrument was delivered. The Clarks, more hesitant, offered to try a 15-inch lens, which they would donate to Mississippi if it were inferior to the Harvard glass. They compromised—the Clarks tackled an $18\frac{1}{2}$-inch objective but would take no money until the instrument was finished.[67]

The glass blanks, ordered from Chance Bros., arrived within a few months. The Clarks, meanwhile, had moved to premises large enough to accommodate this work. By the summer of 1861 the achromatic lens was near enough to completion for the Clarks to ask Barnard to come to Cambridgeport to see it, but the escalation of the war between the states put an end to commerce between Mississippi and Massachusetts, and the Clarks were stuck with a lens insured for $1000. In early January 1863 representatives of the newly formed, and wealthy, Chicago

---

[65] See letters from William R. Dawes to George Knott (published in *Observatory*, vol. 33 [1910], pp. 343-359, 383-398, 419-431, 473-478).

[66] Mable Sterns, *Directory of Astronomical Observatories in the United States*, op cit., p. 116.

[67] John Fulton, *Memoirs of Frederick A. P. Barnard* (New York, 1896), pp. 244-246.

Astronomical Society went to Cambridgeport to buy the lens. The astronomers at Harvard College Observatory, also eager to obtain this lens, had already begun a campaign for the necessary funds; they accelerated their efforts when they learned of the Chicagoans' intent. The Chicagoans arrived just after Alvan Clark had set out to close the deal with George P. Bond; they ran after him, so the story goes, money in hand. It is related that Clark, desirous of keeping the lens in Cambridge, suggested they purchase the Harvard 15-inch instead. The Chicagoans' ready money won out. They bought the lens for $11,187, slightly less than the Harvard 15-inch Munich object-glass had cost.[68]

The Clarks made a German style equatorial mount for the 18½-inch lens, which brought the total cost of the instrument to $18,187. They erected it in a tower of the old University of Chicago (q.v.) in the spring of 1866. That June Alvan Clark received an honorary M.A. from Chicago.[69] After twenty years of fires and financial panics, and the demise of the University, the telescope was moved to Northwestern University (q.v.). In 1911 the lens was remounted by Warner & Swasey. The original mount and wooden tube are now on exhibit at the Adler Planetarium in Chicago.

F. J. DEL CORRAL, astronomer at the Hathorn Observatory at Saratoga Springs, New York, observed the planets with a 6-inch Clark refracting telescope of 1883.[70]

In the Chamberlin Observatory at the UNIVERSITY OF DENVER there is a 20-inch lens figured by Alvan Graham Clark and mounted by Fauth & Co. by 1894. Patterned after the experimental 13-inch Boyden telescope of the Harvard College Observatory (q.v.), the Denver instrument can be used for either visual or photographic work. For the latter, the crown lens of the objective is reversed from its visual position and separated from the flint lens by several inches.[71]

The McKim Observatory of DE PAUW UNIVERSITY in Greencastle, Indiana, was opened in 1885. The equatorial telescope, mounted by

---

[68] *Annals, Dearborn Observatory of Northwestern University*, vol. 1 (1915), introduction.
[69] *University of Chicago Catalogue* (1866).
[70] William H. Knight, "Some Telescopes in the United States," *Sidereal Messenger*, vol. 10 (1891), pp. 398–399. See also *Sidereal Messenger*, vol. 10 (1891), p. 152.
[71] H. A. Howe, "The 20-inch Equatorial of the Chamberlin Observatory," *Astronomy and Astrophysics*, vol. 13 (1894), pp. 709–714.

Figure 15.—Grinding lenses in the Alvan Clark & Sons factory. Although the tools are rotated by means of steam power, the objective lenses are held and moved by hand. From cover of *Scientific American*, 24 September 1887.

Warner & Swasey, has a Clark object glass of 9.53 inches aperture and 12 feet focus.[72]

DOANE COLLEGE in Crete, Nebraska, acquired an 8-inch Clark objective, which was equatorially mounted by Professor Godwin Swezey in 1884.[73]

There is ample evidence that the Clarks made common as well as major astronomical instruments. M. S. DOWLING of Leslie, Michigan, for instance, owned a 3-inch Clark refractor, which he advertised to sell in 1883.[74]

---

[72] W. V. Brown, "McKim Observatory," *Sidereal Messenger*, vol. 4 (1885), pp. 305–307.

[73] "Boswell Observatory," *Sidereal Messenger*, vol. 3 (1884), p. 190.

[74] Advertisement in *Sidereal Messenger*, vol. 1 (1883), front cover of issue # 9.

FIGURE 16.—Polishing lenses in the Alvan Clark & Sons factory. From cover of *Scientific American*, 24 September 1887.

HENRY DRAPER although an outstanding instrument maker himself, purchased several pieces of apparatus from the Clarks. A Clark siderostat on the roof, and secondary mirror, directed sunlight to all corners of his New York City laboratory.[75] This laboratory was used primarily for photographing and studying the spectra of the elements. Around 1879, to verify the existence of oxygen emission lines in the solar spectrum, Draper ordered from the Clarks a spectroscope which would "give the dispersion of twenty heavy flint prisms and . . . bear high magnifying power".[76]

Draper used both reflecting and refracting telescopes at his observatory at Hastings-on-Hudson. He made the reflectors himself, but relied on the Clarks for the refractors. In 1875 he bought a Clark telescope of 12 inches aperture. Alvan Clark thought this objective one of the best he

---

[75] George F. Barker, "Memoir of Henry Draper," *Biographical Memoirs, National Academy of Sciences*, vol. 3 (1895), pp. 113–114.

[76] Henry Draper, "On the Coincidence of the Bright Lines of the Oxygen Spectrum with Bright Lines in the Solar Spectrum," *American Journal of Science*, vol. 18 (1879), p. 268.

had ever seen; while testing it Alvan Graham had discovered the companion of 102 Herculis and the elongation of the principal star of ξ Sagittae.[77] Draper mounted this telescope on the same equatorial stand with his 28-inch silver-on-glass reflector and compared the performances of the two instruments; the photographs taken with each were of about equal clarity. In 1880 this refractor was turned in on a newer model; it was subsequently sold to the Lick Observatory (q.v.) where, in the hands of E. E. Barnard, it yielded exquisite photographs of comets and nebulae.

The newer model was an 11-inch triplet—a visual achromat with a photographic correcting lens—which the Clarks had made for the Lisbon Observatory (q.v.). Its equatorial stand and driving clock were made by Draper himself. With this instrument, on the evening of 30 September 1880, Draper took the first successful nebular photographs. His pictures of the bright part of the Orion nebula, in the vicinity of the trapezium, showed "the mottled appearance of this region distinctly." [78] After Draper's death his widow made the 11-inch available to the astronomers at Harvard (q.v.) as part of the Draper Memorial study of stellar spectra.

During the solar eclipse of 29 July 1878 Henry Draper photographed the corona with a telescope of 5 inches aperture and 78 inches focal length, corrected for photography by the Clarks.[79]

By 1863 the DUDLEY OBSERVATORY, in Albany, New York, had a Clark comet seeker of 4 inches aperture and 42 inches focal length. It was equatorially mounted and supplied with right ascension and declination circles of 5 inches diameter.[80]

When Lewis Boss was appointed director of the Dudley Observatory in 1876 he found most of the instruments in need of repair. As soon as possible, therefore, he engaged the Clarks to bring the apparatus to a "state of efficiency." In his annual report of 1877 Boss had noted that it was "well known that the thirteen-inch equatorial by the late Henry Fitz

---

[77] Edward S. Holden, Obituary of Alvan Clark, *San Francisco Daily Examiner*, 20 August 1887, p. 1. Alvan Clark to Edward S. Holden, 8 March 1876 (letter in Lick Observatory Archives).

[78] Henry Draper, "Photographs of the Nebula in Orion," *American Journal of Science*, vol. 20 (1880), p. 433.

[79] Henry Draper, "The Total Solar Eclipse of July 29th, 1878," *Journal, Franklin Institute*, vol. 106 (1878), pp. 217–220.

[80] *Report of the Astronomer in Charge of the Dudley Observatory* (1863), pp. 4, 10. See also *Annals, Dudley Observatory*, vol. 1 (1866), p. 24.

FIGURE 17.—Machine used by Alvan Clark & Sons for polishing the 36-inch objective for the Lick telescope. While the lens is evenly rotated, a reciprocating motion is imparted to the tool by the two horizontal arms. A grooved pitch lap lies on the floor in front of the machine. From cover of *Scientific American,* 24 September 1887.

of New York has never been regarded as an instrument well adapted for astronomical research. This is due not only to original defects in its optical parts, but also to faulty construction of its mechanical arrangements and the lack of suitable subsidiary apparatus." When the Clarks examined the telescope they found it less faulty than its reputation; they thus limited their work to correcting the chromatic aberration as much as the unusual thinness of the lenses would permit, and making a few additions and improvements to the battery of eyepieces and micrometers. The Clarks also refigured the 6⅜-inch objective of the transit circle so that Boss found it "all that could be desired." And they made a chronograph in which the rotation of the cylinder was governed by a rotary pendulum.[81]

By 1888 FRANCIS G. DU PONT in Wilmington, Delaware, had a private observatory housing a 12-inch aperture reflecting telescope by Brashear, and a refracting telescope with a Brashear equatorial mount and a 4½-inch aperture objective lens figured by the Clarks.[82]

The B. M. C. DURFEE HIGH SCHOOL in Fall River, Massachusetts, has a telescope of 1888. The 8-inch achromatic objective was made by the Clarks, and the clock-driven equatorial mount was made by Warner & Swasey.[83]

EASTERN MICHIGAN UNIVERSITY, a state normal school, has been using a portable 4-inch Clark equatorial refractor since at least 1878. In that year James C. Watson borrowed it for use during the solar eclipse, and with it he "discovered" two new planets between Mercury and the sun.[84]

D. W. EDGECOMB of Newington, Connecticut, owned a 9.4-inch aperture equatorial refractor made by the Clarks in 1874. He claimed that this telescope, which he used for observations of the moon, planets, and

---

[81] *Report of the Astronomer in Charge of the Dudley Observatory* (1877), pp. 6-7. See also ibid. (1878), pp. 1-3.

[82] William C. Winlock, "Progress of Astronomy for 1887, 1888," *Annual Report . . . Smithsonian Institution . . . 1888*, p. 186.

[83] William H. Knight, "Some Telescopes in the United States," op. cit., pp. 394-395.

[84] *U.S. Naval Observatory Reports of the Total Solar Eclipses of July 29, 1879 and January 11, 1880* (Washington, D.C., 1880), p. 117.

FIGURE 18.—Improved telescope eyepiece which permits easy access to the lenses for cleaning and repair, patented by Alvan Clark on 11 November 1851. Drawing from patent papers.

double stars, showed him objects "generally considered tests for 12 inches." [85] In 1880 the instrument passed to Charles P. Howard (q.v)

As pleased as he was with this instrument, Edgecomb was even more enthusiastic about binocular telescopes. His largest pair of binoculars—with prismatic eyepieces and twin objectives measuring $6\frac{5}{16}$ inches each—were made at the Clark establishment in 1894. The prisms, of exceptionally pure glass, were well figured. The focal lengths of the objectives—94 inches—did not differ by more than a hundredth of an inch; the focal lengths of a pair of 5-inch objectives, figured for Edgecomb by Sir Howard Grubb, however, differed by ten times this amount.[86] According to Edgecomb, Alvan Graham took great pride in these binoculars, but was reluctant to give credit where it was due. Writing after the death of Alvan Graham, Edgecomb noted that "the whole work, I

---

[85] Edward S. Holden, "Astronomy," *Annual Record of Science and Industry* (1877), p. 45.

[86] D. W. Edgecomb, "Reflectors," *Popular Astronomy*, vol. 2 (1894–1895), p. 370.

presume it is proper for me to now say, of objectives and prisms, was [Lundin's] and the skill was his." [87]

FRIEDRICH WILHELM RUDOLPH ENGELMANN added an 8-inch Clark glass, equatorially mounted by the Repsolds, to his private observatory in Leipzig in 1881 [88]

WENTWORTH ERCK installed a 7½-inch Clark objective in his observatory at Sherrington, Bray, Ireland, in 1873.[89] This lens, figured twenty years previously, was the first that Clark had sold to Dawes (q.v.). The glass was "so full of bubbles that to one versed in such matters it might have appeared almost worthless for the delicate purpose for which it was intended"; [90] nevertheless, its definition was so exquisite that Erck was able to see Deimos, shortly after its discovery in 1877, even though this outer satellite of Mars was invisible to many larger and more suitably located instruments.[91]

MARSHALL DAVIS EWELL, a Chicago lawyer, purchased a 6¼-inch Clark equatorial and set up a small astronomical observatory in 1886.[92]

B. M. FISH, of Hamburgh, New York, had a 7⅓-inch Clark refractor of 1872 with which he searched for comets.[93]

Captain RICHARD S. FLOYD, who as first president of the Lick Trustees was instrumental in the development of the Lick Observatory, had a private astronomical observatory at Kono Tayee, Clear Lake, California. In 1889 Captain Floyd acquired a 5-inch aperture refracting telescope made by the Clarks. Like the 13-inch Boyden telescope at the Harvard College Observatory (q.v.), Floyd's telescope could be used for both visual and photographic work. The visual observations were made with the two lens components nearly in contact, resulting in a focal length of

---

[87] D. W. Edgecomb, "On the Performance of a 6¼-inch Binocular Telescope," *Popular Astronomy*, vol. 10 (1902), pp. 523–531.

[88] "Telescopes," *Encyclopaedia Brittanica* (9th ed., Chicago, 1894), vol. 3, pp. 149–150.

[89] Wentworth Erck, "Description of an Observatory Erected at Sherrington, Bray," *Observatory*, vol. 1 (1877), pp. 135–137.

[90] *Monthly Notices, Royal Astronomical Society*, vol. 51 (1891), pp. 194–196.

[91] Asaph Hall, *Observations and Orbits of the Satellites of Mars* (Washington, D.C., 1878), p. 34.

[92] *Sidereal Messenger*, vol. 5 (1886), p. 287.

[93] William H. Knight, "Some Telescopes in the United States," op. cit., pp. 396–397.

77 inches. For the photographic correction the crown glass component was reversed and separated from the flint component by 1.7 inches; the photographic focal length was only 65.6 inches.[94]

In 1895, five years after Floyd's death, his daughter donated the 5-inch telescope to the Lick Observatory.[95] Although apparently never mounted on Mount Hamilton, the Floyd telescope was taken on several solar eclipse expeditions. At the Floyd's first fair weather eclipse—that of 22 January 1898—it was used for photographing the general features of the corona; according to W. W. Campbell, it was "a splendid instrument for the purpose." [96]

The astronomical observatory of FRANKLIN AND MARSHALL COLLEGE in Lancaster, Pennsylvania, was built and furnished in 1885-86. The main dome enclosed a Clark 11-inch objective equatorially mounted by Repsold & Sons of Hamburg, Germany.[97]

DAVE GARROWAY, of New York City, bought the 9½-inch aperture Clark-Warner & Swasey equatorial from Beloit College (q.v.) in 1967.

WINTHROP S. GILMAN, JR., a banker living in Palisades, Rockland Co., New York, enthusiastically used several Clark instruments. He observed the 1869 total solar eclipse with his 4-inch Clark achromat equipped with a ruled micrometer glass which gave position angles for every 45°; the mounting was Clark's "usual portable equatorial, without circles or clockwork." [98] Some of Gilman's other observations were reported in letters to the director of the Harvard College Observatory. These letters recorded "sunspots as seen August 14, 1868 in a 9-inch glass of Alvan Clark's make," as well as various observations with Tasker Marvin's 5-inch (q.v.) and Jacob Campbell's 12-inch (q.v.) Clark lenses.[99]

---

[94] *Sidereal Messenger*, vol. 8 (1889), p. 92.

[95] Edward S. Holden, "Gift to the Lick Observatory—The Floyd Photographic Telescope," *Publications, Astronomical Society of the Pacific*, vol. 7 (1895), p. 339.

[96] W. W. Campbell, "A General Account of the Lick Observatory—Crocker Eclipse Expedition to India," *Publications, Astronomical Society of the Pacific*, vol. 10 (1898), p. 130.

[97] *The [Franklin and Marshall] College Student*, vol. 5 (1884), pp. 21-23; and ibid., vol. 6 (1886), pp. 69, 74.

[98] *U.S. Naval Observatory Reports on Observations of the Total Eclipse of the Sun, August 7, 1869*, p. 173.

[99] See letters from Winthrop S. Gilman to Joseph Winlock ca. 1868-1869 (in Observatory Papers, Harvard University Archives).

Figure 19.—Spectroscope with 6½ prisms twice-traversed, made by Alvan Clark & Sons for Charles A. Young at Dartmouth College. Courtesy Dartmouth College Archives.

Captain CHARLES GOODALL of San Francisco, a donor of the observatory and apparatus for the University of the Pacific (q.v.), had an excellent 5-inch Clark achromat of his own by 1888.[100]

GRAND RAPIDS HIGH SCHOOL in Grand Rapids, Michigan, loaned its 3¾-inch Clark achromat to James Craig Watson for observations of the 1870 solar eclipse. Watson, who used the telescope under a variety of circumstances, was convinced its optical properties could not be excelled by any other instrument of that size.[101]

ROBERT GREEN, of Mexico, Missouri, owned a 4-inch aperture refracting telescope inscribed "Alvan Clark & Sons, Cambridgeport, Mass., 1885." His widow gave this instrument to the Central Missouri Amateur Astronomers (q.v.).[101a]

GRINNELL COLLEGE acquired an equatorial refractor in 1888. The Clarks made the 8-inch objective and a battery of four eyepieces magni-

---

[100] Edward S. Holden, *Handbook of the Lick Observatory* (San Francisco, 1888), p. 125.
[101] *U.S. Coast Guard Survey Report* (1870), p. 130.
[101a] Information from W. C. Shewmon of Central Missouri Amateur Astronomers.

fying from 100 to 800 diameters. The mount was provided by the Rev. H. G. Sedgewick of Davenport, Iowa.[102]

PHOEBE HAAS, of Philadelphia, received a 4-inch Clark refractor from William Tyler Olcott (q.v.), and then passed it on to the American Association of Variable Star Observers (q.v.).

The first of GEORGE ELLERY HALE'S great telescopes was a 4-inch Clark refractor given to him by his father in time for observations of the 1882 Transit of Venus. When the Kenwood Physical Observatory was furnished, nine years later, the new Warner & Swasey equatorial mount was made strong enough to carry the small Clark telescope as well as the 12-inch Brashear refractor.[103]

The HARTFORD PUBLIC HIGH SCHOOL in Connecticut has had an excellent 9½-inch Clark objective, equatorially mounted by Warner & Swasey, since 1884.[104] A Hartmann test applied to the objective in 1930 gave a T-value of only 0.18. An accompanying diagram shows, in the author's words, the astigmatism, "or rather, the lack of astigmatism" of the lens, in thousandths of an inch.[105]

HARVARD COLLEGE OBSERVATORY, within a few miles of the Cambridgeport factory, was advantageously located, and there is ample evidence that the Harvard astronomers relied on the Clarks for many of their astronomical instruments. The observatory's annual reports, for instance, refer to work done by the Clarks: a new comet seeker, a new solar eyepiece, cleaning lenses, devising new modes of illumination for micrometer wires, etc. The report for 1892 went so far as to note the death of George Bassett Clark and to pay tribute to his "genius for mechanical devices, indomitable perseverance, and devotion to the interests of the observatory." [106]

---

[102] *Proceedings on the Reception of the Jacob Haish Telescope, June 19, 1888* (Grinnell, Iowa, 1888), p. 1.
[103] George Ellery Hale, "The Kenwood Physical Observatory," *Sidereal Messenger*, vol. 10 (1891), pp. 321–323.
[104] "The Hartford High School Telescope," *Sidereal Messenger*, vol. 3 (1884), pp. 147–149.
[105] E. Harold Coburn, Quality of H.P.H.S. Telescope Objective, Hartmann Test (manuscript copy at Hartford Public High School).
[106] *Annual Report, Harvard College Observatory*, vol. 47 (1892), p. 3.

The first recorded job the Clarks did for Harvard was early in 1857; they made a glass prism and were paid $100.[107] Many years later the Clarks were involved in another Harvard prism incident. J. Trowbridge, while a green, young instructor at Harvard, borrowed two valuable prisms from the observatory and succeeded in chipping them; Joseph Winlock, the director, replied only, "Oh, I always intended to get Alvan Clark to reduce the size of these prisms, and he would have had to chip off these edges." [108]

Later that year the Clarks played a key role in the first successful stellar photography. The photographic experiments, made by George Bond and John Whipple in 1850, had been hindered as much by the irregularity of the Munich clock drive as by the slowness of the daguerreotype process. In a letter to the Royal Astronomical Society, seven years later, Bond attributed their success to the collodion process, and to the new spring-governor controlled clock drive "which has been adapted to the great telescope by those excellent mechanicians, Messrs. George Clark and Alvan Clark, jun., of East Cambridge, assisted by their father, Mr. Alvan Clark, sen." [109] The first photograph, taken on 27 April 1857, showed Mizar, its fourth-magnitude companion, and Alcor; the distances between the images of these stars were measured the next day by Alvan Clark with a reading microscope.[110] Inspired by Bond's success, at least two notable astronomers—Dawes and Rutherfurd (qq.v.)—were led to order equatorial mounts with spring-governors from the Clarks.

Joseph Winlock became director of the Harvard observatory in 1866 and began immediately to improve and increase the equipment. In 1869 the Clarks completed the West Equatorial, a refractor of 5¼ inches aperture, and 7½ feet focal length, equipped with divided circles, spring-governor driving clock, and spectroscope. The spectroscope, similar to that made for Charles Young at Dartmouth (q.v.), consisted of five prisms of dense Munich glass through which the light passed twice.[111]

---

[107] *Harvard College Papers*, vol. 24, 2nd ser. (1857), p. 119.

[108] Quoted in J. Trowbridge, "Scientific Cambridge," in A. Gilman (ed.), *The Cambridge of Eighteen Hundred and Ninety Six* (Cambridge, Mass., 1896), p. 76.

[109] "Letter From Mr. Bond, Director of the Observatory, Cambridge, U.S., to the Secretary," *Monthly Notices, Royal Astronomical Society*, vol. 17 (1856-1857), pp. 230–232.

[110] Ibid.

[111] *Annals, Harvard College Observatory*, vol. 8 (1877), p. 35.

This telescope was frequently taken on solar eclipse expeditions: as mentioned above, George Clark and John Whipple used it photographically at Shelbyville in 1869. Winlock was anxious to show, from these 1869 photographs, that pictures taken in the principal focus of a telescope are less distorted than those taken after enlargement by the eyepiece. To this end he directed the Clarks to construct a micrometer to measure the distance viewed under a microscope between the centers of the photographic images of the sun and moon. The results indicated the great accuracy of this method of observation,[112] and most photographs of the 1870 eclipse were taken in the principal focus.

Shortly after the eclipse Winlock designed, and the Clarks constructed, a fixed horizontal telescope for solar photography. This instrument was so successful that a modified version of it was chosen for the official American photographic observations of the Transits of Venus (q.v.) of 1874 and 1882. Sunlight was reflected to the objective by a flat mirror. Rather than using a heliostat, whose irregular motion would cause some distortion, Winlock manually adjusted the mirror before each exposure. A fast moving shutter, to give as short an exposure to the glaring sun as possible, was the most difficult part of this camera. The Clarks figured two simple 4-inch objectives: one visual, of 40 feet focus; and one photographic, of $32\frac{1}{2}$ feet focus. Winlock found, however, that they gave equally good images. Their extreme focal lengths insured large solar images and, like the aerial telescopes of the 17th century, lessened the effects of the non-achromatism of the lens.[113]

The meridian circle Winlock designed for the Harvard Observatory was mounted in 1870. The metal work was done by Troughton and Simms of London, while the optical parts were figured by the Clarks. This included the objectives of the principal telescope—$8\frac{1}{4}$ inches aperture, and 9 feet 4 inches focus—and of the two collimators—the same focus, but slightly smaller diameter. Fourteen years later William A. Rogers measured the errors of graduation of this meridian circle by means of electromagnetic clamps. George Clark designed and supervised construction of the equipment and he participated in the measurements.[114]

---

[112] See Winlock's report on the solar eclipse of 1869 in *U.S. Coast Survey Report* (1869), pp. 124–125.

[113] *Annals, Harvard College Observatory*, vol. 8 (1877), pp. 35–42.

[114] William A. Rogers, "On the Original Graduation of the Harvard College Meridian Circle *in situ*," *Sidereal Messenger*, vol. 3 (1884), pp. 306–311.

Edward C. Pickering, who succeeded Winlock as director of the Harvard Observatory in 1875, was especially interested in the new sciences of stellar photometry and spectroscopy. As very few instruments had yet been developed for these aspects of astronomy, the Clarks were called upon to help design as well as construct much of the apparatus for Pickering's studies. For the photometry the most important of the new instruments were the two meridian photometers, with which the light of any star near the zenith could be compared with that of a reference star near the north pole.[115] Each photometer employed two similar and horizontal objectives with a mirror—or prism—to reflect the starlight through them; the two images were analyzed by means of a Nicol prism. The smaller photometer, with 4-cm. objectives, was completed by 1879; while the larger, with 10.5-cm. objectives, was in use three years later.

Harvard at this time acquired several smaller, special purpose photometers as well. In their *Annual Report* for 1878 they noted that great credit was due to George Clark, "without whose ingenuity and skill their construction would have been attended with great difficulty." Photometer M, typical of these new instruments, was a double-image micrometer which was used for measuring the positions and distances, as well as equalizing the brightness of the components of double stars.[116]

All of the large instruments used in Harvard's Henry Draper Memorial Study of stellar spectra were made by the Clarks, under the special supervision of George Clark. In 1885, with a grant from the Bache fund of the National Academy of Sciences, Pickering obtained a telescope specially adapted for stellar photography. He bought a Voightlander portrait lens of 8 inches aperture and 45 inches focus; this doublet, or combination of two sets of crown and flint components, has a low f/number, and thus is photographically fast. This lens system was then given to the Clarks for correction and mounting. The correction involved lengthening the focus by about 10 centimeters so that the scale of the photographs would be the same as that of various other star charts. The telescope was held in an equatorial fork mount equipped with a Bond spring-governor drive.[117]

With an 8-inch prism over the objective this telescope was used for

---

[115] *Annals, Harvard College Observatory*, vol. 14, pt. 1 (1884), and vol. 23 (1890).

[116] *Annual Report, Harvard College Observatory*, vol. 33 (1878), pp. 5–6.

[117] Edward C. Pickering, "Stellar Photography," *Memoirs, American Academy of Arts and Sciences*, vol. 11 (1888), p. 184.

photographing stellar spectra, and the results formed the basis of the *Draper Catalogue*. In 1889 the Bache telescope was sent to Harvard's observatory in Peru for studies of the southern stars. To continue the unfinished studies in Cambridge, Henry Draper's widow provided money for another photographic instrument. This telescope, also made by the Clarks, differed from the Bache telescope only in its slightly longer focus and its German rather than forked equatorial mount.[118] In 1947 the Draper doublet was transferred to an observatory near Torun, Poland, the birthplace of Copernicus.

For detailed study of the spectra of the brighter stars the Harvard astronomers used the 11-inch photographic telescope which had belonged to Henry Draper (q.v.). A battery of four objective prisms, each with a clear aperture of 11 inches, was constructed by the Clarks for use with this telescope. When all four prisms were used simultaneously the spectral images were over 4 inches long, and, in some cases, showed as many as 500 lines.[119] In 1947 Harvard sent this telescope on a long-term loan to the Sun Yat-sen University Observatory, in Canton, China.[120]

Several instruments used in the high altitude Boyden investigations were from the workshops of Alvan Clark & Sons. Perhaps the most novel of these was the visual-photographic refracting telescope. This lens combination could be focused for either the yellow or blue rays by altering the direction and position of the crown glass component (see above, p. 000). Early in 1887 the Clarks made a smaller, experimental model which proved so successful that they received an order for the 13-inch objective. According to E. C. Pickering, both the visual and photographic images formed by these instruments were excellent.[121] The mount which held the 13-inch refractor also held a Clark-made photographic doublet of 8 inches aperture and 11 feet focal length. These telescopes were used at Willows, California, for observations and photographs of the solar eclipse of 1 January 1889; the following year they were sent to Harvard's observing station at Arequipa, Peru. Among the other Boyden instruments made

---

[118] *Annals, Harvard College Observatory*, vol. 26, pt. 1 (1891), p. xiii.

[119] Ibid., pp. xv–xvi.

[120] Dorrit Hoffleit, "A Famous Old Telescope Goes to China," *Sky and Telescope*, vol. 7 (1947), pp. 8-9.

[121] Edward C. Pickering, "New Form of Construction of Object-Glasses Intended for Stellar Photography," *Nature*, vol. 36 (1887), p. 562.

by the Clarks were a 12-inch visual refractor and a 10-inch refractor equipped with a photographic lens.

The last major instrument the Clarks made for Harvard was the 24 inch Bruce photographic telescope. Like the 8-inch Bache camera, this had a doublet lens of short focus and an objective prism. It was compact and fast and covered a wide field. For half a century this telescope was used to photograph the southern skies; among its triumphs were the plates of the Magellanic Clouds on which the luminosity pattern of the Cepheid variables was discovered.

Since the Clarks made apparatus for physical as well as astronomical researches, it is not surprising to find Clark apparatus in departments of physics and chemistry. By 1878 the Rumford Cabinet at HARVARD UNIVERSITY had a Clark objective lens of $6\frac{1}{2}$ centimeters aperture and 1.6 meters focus, which was used as a collimator for a spectroscope.[122]

Josiah Parsons Cooke, Jr., Professor of Chemistry and Mineralogy at Harvard University, used at least two spectroscopes made by the Clarks during the early 1860's. The first, probably a simple chemical spectroscope, was in use by 1862. Since large and perfect pieces of optical glass were difficult to obtain, opticians tried various other methods of producing a prism. For the small spectroscope the Clarks made a prism about 4 inches on each face, built up of several pieces of glass cemented together with canada balsam.[122a] The Clarks also made for Cooke some prisms of hollow glass filled with carbon bisulphide. Although this liquid smelled foul, and its index of refraction changed greatly with even slight temperature changes, it gave a large dispersion. The prism frames were composed of pieces of plate glass cemented together with a mixture of glue and honey. To the outside surfaces of these were held, with castor oil, carefully figured and polished plates of glass. The largest prisms of this sort had faces about 5 inches long by 3 inches high and held nearly a pint of liquid apiece.[122b]

---

[122] "List of Apparatus Relating to Heat, Light, Electricity, Magnetism, and Sound, Available for Scientific Researches Involving Accurate Measurements." *Annual Report . . . Smithsonian Institution . . . 1878*, p. 431.

[122a] Letter to the editor from Josiah P. Cooke, Jr., *American Journal of Science*, vol. 34 (1862), pp. 299–300.

[122b] Note by Benjamin Silliman, Jr., in *American Journal of Science*, vol. 35 (1863), p. 408. See also, Josiah P. Cooke, Jr., "An Improved Spectroscope," *American Journal of Science*, vol. 36 (1863), pp. 256–257.

The second spectroscope, built in 1863, was the largest and most powerful yet made. The spectrum was produced by a train of nine carbon bisulphide prisms; and an alternate set of solid glass prisms—not figured by the Clarks—was also provided. The instrument incorporated a new method, devised by Cooke and executed by George Clark, of adjusting the prisms for the angle of minimum deviation for any ray. At the center of the spectroscope was a truncated iron cone, against which the backs of the prisms rested: by raising or lowering the cone the prisms were equally spread or drawn toward the center. For measuring the relative positions of the spectral lines, the rim of the spectroscope was made of silver and graduated to 10 seconds of arc.[122c]

In 1902 CHARLES S. HASTINGS, the theoretical as well as practical optician, advertised to sell cheaply a 4-inch refracting telescope "made by Alvan Clark." [123]

The Fitz equatorial refractor at HAVERFORD COLLEGE was reworked by the Clarks in 1880. They corrected and repolished the 8¼-inch objective, made a new micrometer, and added a rotary pendulum regulator to the driving clock. Even with these improvements, however, the telescope was not capable of very accurate work, and a new instrument was ordered from the Clarks. The 10-inch refractor, erected in 1884, had an equatorial mount equipped with clock drive and graduated circles.[124] This instrument was reconditioned after fifty years' service and is still in use.

HEBRON ACADEMY in Hebron, Maine, owns a 6.1-inch aperture Clark refractor dated 1859. The tube, originally mounted on a tripod, is of brass coated with nickel. The telescope has recently been remounted and reactivated.[125]

ANDREW P. HENKEL, a physician in Cincinnati, Ohio, owned an 8¼-inch aperture refracting telescope made by the Clarks in 1880. The equatorial mount was of the usual Clark form; the polar axis was sup-

---

[122c] Josiah P. Cooke, Jr., "On the Construction of a Spectroscope with a Number of Prisms, by Which the Angle of Minimum Deviation for Any Ray May Be Accurately Measured and Its Position in the Solar Spectrum Determined," *American Journal of Science*, vol. 40 (1865), pp. 305–313.

[123] Advertisement in *Popular Astronomy*, vol. 10 (1902).

[124] *Haverford College Observatory* (1884) (pamphlet in Haverford College Library).

[125] Private correspondence with Jerald R. Twitchell, chairman of Hebron Academy science department.

ported by either end of an open arc frame which enclosed the right ascension setting circle and many of the clock drive gears. The two setting circles were finely engraved on silver and each was read under two magnifying glasses. This telescope is now owned and used by the Cincinnati Astronomical Society (q.v.).[126]

JOHN R. HOOPER of Baltimore had a 5-inch Clark refractor of 1866 which was equatorially mounted and equipped with graduated circles and a clock drive.[127]

CHARLES P. HOWARD acquired the 9.4-inch Clark telescope originally owned by D. W. Edgecomb (q.v.) in 1880, and mounted it in his private observatory in Hartford, Connecticut. Howard observed several solar eclipses, including that of 28 May 1900, with the 2-inch aperture finder of this instrument.[128] The large telescope was subsequently sold to someone in the Midwest, and the finder was sold to Roland Bourne (q.v.).

The early spectroscopic discoveries of WILLIAM HUGGINS were made with starlight gathered by an 8-inch Clark object glass.[129] Huggins had bought this lens from Dawes (q.v.) in 1858, had it remounted by Thomas Cooke, and installed it in his observatory at Upper Tulse Hill, a suburb of London. Ten years later, when a much larger instrument was being made for him by Howard Grubb, Huggins sold the 8-inch to a Mr. Corbett.[130]

When JEFFERSON COLLEGE, at Canonsburg, Pennsylvania, was in the market for a telescope they applied to Albert Hopkins of Williams College for advice.[131] Hopkins recommended an instrument similar to the one at Williams (q.v.) and by 1859 students at Jefferson College were using a refracting telescope with a 7½-inch Clark lens and an equatorial

---

[126] Information from Roland E. Johnson, President, Cincinnati Astronomical Society.

[127] William H. Knight, "Some Telescopes in the United States," op. cit., pp. 394-395.

[128] Charles P. Howard, *Total Eclipse of the Sun, May 28, 1900* (Hartford, Conn.: R. S. Peck & Co., 1900[?]), pp. 14-16.

[129] William Huggins, and W. A. Miller, "On the Spectra of Some of the Fixed Stars," *Philosophical Transactions, Royal Society*, vol. 154 (1864), p. 415.

[130] Mary Needham, A Gentleman and a Scholar and His Lady Wife (typescript copy at Wellesley College Observatory).

[131] Albert Hopkins to William C. Bond, 7 February 1859 (letter in Bond Papers, Harvard University Archives).

mount by Phelps.¹³² Six years later Jefferson College merged with the nearby Washington College to form Washington and Jefferson College (q.v.), and soon thereafter the scientific work and equipment were moved to the campus at Washington, Pennsylvania.

The COLLÈGE DES JÉSUITES in Quebec is the present owner of the 8-inch Clark equatorial refractor originally mounted in the Canadian observatory on Bonner's Hill (q.v.).

JOLIET HIGH SCHOOL in Joliet, Illinois, has had a 4½-inch Clark equatorial refractor since at least 1891.¹³³

In 1855 Baron de Rottenburg ordered a Clark telescope for subscribers in Kingston, Canada West. This 6¼-inch refractor, on a plain equatorial mount, sold for $850.¹³⁴ Six years later the KINGSTON OBSERVATORY, with its apparatus, was transferred to Queen's University (q.v.) in Kingston, where it is yet. The Clark telescope has only recently been retired.

In 1857 GEORGE KNOTT, of Sussex, England, wrote to W. R. Dawes (q.v.) in search of a suitable telescope for his private use. Dawes' reply was an offer of the 7⅓-inch refractor Clark was then constructing for him, together with his Munich equatorial mount. Two years later, when Alvan Clark brought him a replacement, Dawes sold the 7⅓-inch to Knott.¹³⁵

KNOX COLLEGE in Galesburg, Illinois, acquired the 6-inch Clark equatorial, along with other instruments from Edgar Larkin's observatory (q.v.) in 1888. This telescope was used, mainly for instruction, until the college observatory was torn down a few years ago.¹³⁶

An 8-inch Clark objective equatorially mounted by Repsold, was erected at the KÖNIGLICHE UNIVERSITÄTS-STERNWARTE in Breslau, Germany, in 1898.¹³⁷

---

[132] Elias Loomis, *Practical Astronomy* (New York, 1866), 2nd ed., pp. 496-497.
[133] William H. Knight, "Some Telescopes in the United States," op. cit., pp. 396-397.
[134] Elias Loomis, *Recent Progress of Astronomy* (New York, 1856), p. 391.
[135] See letters from William R. Dawes to George Knott (published in *Observatory*, vol. 33 [1910], pp. 343-359, 383-398, 419-431, 473-478).
[136] Private correspondence with Mrs. Philip Haring, Curator, Knox College.
[137] "Observatory," *Encyclopaedia Brittanica*, vol. 19 (Cambridge, 1911), 11th ed., p. 957.

EDGAR LUCIEN LARKIN, in 1880, equipped his private observatory in New Windsor, Illinois, with a 6-inch Clark equatorial refractor. In 1888 Larkin transferred his instruments to Knox College (q.v.), and he became director of that school's observatory. When Larkin moved to Mount Lowe (q.v.), his instruments remained at Knox.[138]

The Underwood Observatory at LAWRENCE UNIVERSITY in Appleton, Wisconsin, was ready for use in 1892. The 10-inch equatorial refractor and the 4-inch meridian transit had Clark objectives; the mounts were by Sedgewick, undoubtedly the man responsible for the equatorial at Grinnell College (q.v.).[139]

Among the many, and now probably misplaced, bread and butter Clark instruments was a 4-inch equatorial—without driving clock and micrometer—sold to Mr. U. W. LAWTON of Jackson, Michigan, in 1891.[140]

Shortly after its founding, LEHIGH UNIVERSITY at South Bethlehem, Pennsylvania, acquired a 6-inch Clark equatorial refractor, equipped with driving clock and filar micrometer. Although intended primarily for teaching, this instrument was also used for astronomical discovery. Through it, in 1870, Alfred M. Mayer saw a "ruddy, elliptical line" on Jupiter. The only other contemporary observation of the great red spot was made by Gledhill with a Cooke refractor of $9\frac{1}{3}$ inches aperture.[141] The Lehigh objective was remounted during the past quarter century, and the original mount has been lost.[142]

The astronomical observatory of LEIDEN, in the Netherlands, installed a new telescope in 1885: the equatorial mount was built by the Repsolds, and the $10\frac{1}{2}$-inch aperture objective lens was figured by Alvan Clark & Sons.[143]

JOHN M. LEWIS, of Mount Vernon, Ohio, advertised a used portable

---

[138] Private correspondence with Mrs. Philip Haring, Curator, Knox College.
[139] *Sidereal Messenger*, vol. 9 (1890), p. 421.
[140] *Sidereal Messenger*, vol. 10 (1891), p. 522.
[141] Alfred M. Mayer, "Observations on the Planet Jupiter," *Journal, Franklin Institute*, vol. 59 (1870), pp. 136–139; see also vol. 60 (1870), p. 82.
[142] Private correspondence with Ralph N. Van Arnam, Department of Mathematics and Astronomy, Lehigh University.
[143] "Report of the Director of the Leiden Observatory for the Year Ending 15 Sept. 1885," reported in *Observatory*, vol. 9 (1886), p. 138.

transit instrument in 1884. He neglected to state the maker of the mount but emphasized the perfections of the 3-inch Clark objective.[144]

The first observations with the 36-inch LICK equatorial refractor were made on 3 January 1888, fourteen years after James Lick first announced his bequest. It was then the largest instrument of its kind anywhere in the world, and its performance matched its size. The history of this great telescope has often been told. Warner & Swasey made the mount; the Clarks made the 36-inch objective and 33-inch photographic corrector. The contract with the Clarks was not let until 1880. Then there were long delays before Feil could produce satisfactory glass blanks: although the flint disc was soon cast, a successful crown glass was obtained only after another $3\frac{1}{2}$ years and 19 failures and a trip by Alvan Graham to Paris; in 1887 Alvan Graham went to Paris again to negotiate for a better crown disc for the photographic lens as the first crown disc intended for this purpose had been defective and had broken under the Clarks' hands. The visual achromatic combination was sent to Mt. Hamilton in 1886; Alvan Graham accompanied the photographic lens west the following year.

During these fourteen years representatives of the Lick Trust visited the Cambridgeport workshop, first to advise whether the Clarks be given the contract, and later to report on work in progress. Their letters give some intimate pictures of the Clarks' personalities and habits. In 1877 Howard Grubb, who was also eager for the contract, told the trustees that Alvan Clark "had set his heart on making the big object glass."[145] Charles Plum visited the Clarks in 1879 and reported that they were "wholly devoted to the manufacture of instruments for the advancement of science in astronomy, and are continually experimenting for improvements."[146] Eight years later, when the trustees were becoming impatient and the fate of the photographic lens was doubtful, he noted that the Clarks "don't save any time when it is to cost them the difference in money."[147] Although many visitors were impressed by the Clarks' friendly and talkative manners, Simon Newcomb warned the trustees not to expect voluntary progress reports, as "the Clarks never write unless they

---

[144] *Sidereal Messenger*, vol. 3 (1884), pp. 287–288.

[145] Howard Grubb to Richard S. Floyd, 4 September 1877 (letter in Lick Observatory Archives).

[146] Charles Plum to E. Matthews, 3 June 1879 (ibid).

[147] Charles Plum to Lick Trust, 21 May 1887 (ibid.).

have something to say."[148] The truth of Newcomb's comment is confirmed by the existing Clark correspondence, all of which is consistently direct. When Alvan Graham was in California for the installation of the telescope, James Keeler found him "a terrible old blow and grumbler" who insisted that "the only decent thing about the telescope is the object glass, that the dome is worthless and the shutter the same."[149]

Besides the 36-inch, the Clarks worked on several other Lick instruments: the 12-inch equatorial refractor which had belonged to Henry Draper (q.v.); three 6-inch lenses for the telescope and two collimators of the Repsold meridian circle—one of which also served as the object glass of a Warner & Swasey portable equatorial; an excellent 4-inch objective for a Fauth transit instrument; a 4-inch comet seeker which employed a reflecting prism to send the light in one horizontal plane; and a photoheliograph essentially identical to the American Transit of Venus instruments (q.v.).[150] In addition, the Lick Observatory inherited the 5-inch aperture visual/photographic refractor the Clarks had made for Richard S. Floyd (q.v.)

The history of the 12-inch refractor illustrates some of the problems encountered in identifying a well-pirated instrument. Before sending this telescope to California the Clarks, acting on Simon Newcomb's instructions, altered its drive mechanism. They replaced the original spring governor, which they found more reliable in practiced hands, with the more durable rotary pendulum. A few years later, by the time the Lick astronomers had decided in favor of the original regulator, the Clarks had fitted it to the drive of the equatorial they were making for the United States Military Academy (q.v.).[151]

In 1877 the Clarks completed an 11-inch refractor, with a photographic correcting lens, for the LISBON OBSERVATORY. The mount for this was described as stable and elegant.[152] This instrument never reached its Portuguese destination and was sold to Henry Draper (q.v.).

---

[148] Simon Newcomb to Richard S. Floyd, 6 July 1880 (ibid.).

[149] James Keeler to Edward S. Holden, 6 January 1888 (ibid.).

[150] For descriptions of the original Lick instruments see *Publications, Lick Observatory*, vol. 1 (1887).

[151] Alvan Clark & Sons to Richard S. Floyd, 2 August [1883] (letter in Lick Observatory Archives).

[152] Edward S. Holden, "Astronomy," *Annual Record of Science and Industry* (1877), p. 28.

MISS ELMA LOINES gave the Maria Mitchell Observatory (q.v.) an 8-inch refractor, made by the Clarks in 1871, which her father had used in his observatory at Bolton Landing, Lake George, New York.[153]

REV. R. E. LOWE, in England, purchased an 8¼-inch Clark equatorial from Dawes (q.v.) in 1864 and later sold it to J. M. Wilson for use at Rugby School (q.v.). Lowe had Thomas Cooke make a new drive mechanism for the telescope and then wrote sadly: "And perfect as one has been taught to consider Cooke's mountings to be, this Alvan Clarke clockwork was pronounced to take decided precedence over that by which the York maker replaced it."[154]

Thaddeus S. C. Lowe built the LOWE OBSERVATORY on Echo Mountain, at the head of the Mount Lowe cable railway, in 1894. Other attractions on this California mountain included several hotels and pleasure resorts. Lowe invited Lewis Swift (q.v.) to direct the new observatory, and to bring with him his 16-inch Clark equatorial refractor which had been installed in the Warner Observatory (q.v.).[155] The Lowe Observatory was in existence until 1928, when the dome was blown off by an extra strong wind. In 1941 the Pacific Electric Railway Company, which had acquired title to the Mount Lowe properties, sold the telescope to the University of Santa Clara (q.v.).

Alvan Graham Clark worked closely with Percival Lowell in planning and equipping the LOWELL OBSERVATORY. Before the Flagstaff, Arizona, site had been chosen, Lowell and his assistants used portable Clark equatorials of 4 and 6 inches aperture to test atmospheric conditions at other possible sites around the world; the definition of the 6-inch telescope was said to be "unsurpassed in beauty."[156]

The first large telescope erected at Lowell was a composite instrument. The equatorial mount, borrowed from Harvard (q.v.), had been built by the Clarks to carry the 13-inch Boyden photographic refractor. At Lowell it was adapted to carry a refracting telescope at either end of the declination axis: an 18-inch Brashear and a 12-inch Clark. The 12-inch, re-

---

[153] Private correspondence with Dorrit Hoffleit, Director of Maria Mitchell Observatory.

[154] R. E. Lowe to G. M. Seabroke, 14 September 1870 (letter in Rugby School Archives).

[155] C. D. Perrine, "The Lowe Observatory," *Publications, Astronomical Society of the Pacific*, vol. 7 (1897), pp. 47-48.

[156] *Annals, Lowell Observatory*, vol. 1 (1898), historical introduction.

putedly a model for the Lick telescope, was used photographically. This instrument was in use by the summer of 1894.[157]

Hoping to confirm the revelations of the 1894 Martian opposition, Percival Lowell planned to observe the next close approach of Mars under even more favorable conditions. Accordingly, he ordered a 24-inch equatorial refractor from the Clarks. Because of the astronomical deadline this instrument was constructed in less than a year. Nevertheless, this last large and complete telescope built by Alvan Graham was exceptionally perfect. When the lively aspects of the planets seen through it by the Lowell observers, and unconfirmed elsewhere, led E. S. Holden to suggest the Lowell lens was improperly held in its cell, Alvan Graham publicly defended the perfection of the telescope.[158] The glass discs, cast by Mantois, were particularly pure. To figure them the Clark opticians used glass forms—or tools—rather than the customary metal ones.[159] The result was such that, according to Dr. Hartmann himself, the elimination of the residual spherical aberration was more complete than in any lens yet tested.[160] The Burnham-type micrometer showed uniformly bright threads against a perfectly dark field. And the mount was said to be "by far the heaviest and strongest . . . of any telescope of this size in the world."[161]

In 1909 Carl Lundin, then of the Alvan Clark & Sons Corporation, figured a 42-inch Newtonian reflector for Lowell. The mirror was in continuous use for half a century, until it was broken while being converted to a Cassegrain.

Percival Lowell seems to have been a friend as well as a customer of Alvan Graham's. In December 1895 the two men went together to Paris to visit astronomical acquaintances.[162] The following July they went together to Arizona to set up and test the 24-inch telescope. And a week

---

[157] Andrew E. Douglass, "The Lowell Observatory and Its Work," *Popular Astronomy*, vol. 2 (1894–1895), pp. 395–397.

[158] Alvan Graham Clark, letter to the editor, *Science* n.s., vol. 5 (1897), p. 768.

[159] "Another Large Telescope," *Scientific American*, vol. 73 (1895), p. 230.

[160] V. M. Slipher, "The Lowell Observatory," *Publications, Astronomical Society of the Pacific*, vol. 39 (1927), pp. 144–145. See also Gerhard R. Miczaika and William M. Sinton, *Tools of the Astronomer* (Cambridge, Mass., 1961), p. 61.

[161] Thomas J. J. See, "A Sketch of the New 24-inch Refractor of the Lowell Observatory," *Popular Astronomy*, vol. 4 (1896–1897), pp. 297–300.

[162] A. Lawrence Lowell, *Biography of Percival Lowell* (New York, 1935), pp. 92–95.

before his death Alvan Graham accompanied Lowell and his daughter to Amherst to visit mutual friends.[163]

The MARE ISLAND NAVAL OBSERVATORY, in a shipyard in San Francisco, was established by the U.S. Navy for the purpose of rating chronometers and supplying standard time to the West Coast. By 1899 it housed a Clark 5-inch equatorial refractor (#861) mounted on a substantial pier.[164] After the demise of the observatory in 1930 the telescope was probably sent—or returned—to the U.S. Naval Observatory in Washington (q.v.).

In September 1891 E. S. MARTIN, at Wilmington, North Carolina, observed a transit of Jupiter's third satellite with a 5-inch refractor he had just received from Messrs. Alvan Clark & Sons.[165]

TASKER H. MARVIN, of Palisades, New York, had a 5-inch Clark refractor by 1868. His friend and neighbor, Winthrop Gilman (q.v.), was able to see the 16th magnitude companion of 110 Herculis with this splendid glass.[166]

The MASSACHUSETTS INSTITUTE OF TECHNOLOGY, located in Cambridgeport, was within a few blocks of the Clark workshop. By 1878 the M.I.T. department of physics had at least one instrument made by the Clarks—a spectrometer. Like the instruments at Dartmouth and Harvard (qq.v.), this one held five 60° prisms and one 30° prism silvered on its rear surface; and it used the same telescope for both the collimator and viewing telescope.[167]

ROBERT WHITE MCFARLAND, a mathematician at Columbus, Ohio, had a 5-inch Clark equatorial by 1880.[168]

---

[163] "Alvan Graham Clark," *The Cambridge Chronicle* 12, June 1897.
[164] Everett Hayden, "Brief Account of the [Mare Island] Observatory," *Publications, Astronomical Society of the Pacific*, vol. 11 (1899), p. 112.
[165] *Sidereal Messenger*, vol. 10 (1891), p. 431.
[166] Winthrop S. Gilman to Joseph Winlock, 18 July 1868 (letter in Observatory Papers, Harvard University Archives).
[167] "List of Apparatus Relating to Heat, Light, Electricity, Magnetism, and Sound, Available for Scientific Researches Involving Accurate Measurement," *Annual Report . . . Smithsonian Institution . . . 1878*, p. 437. See also Wolcott Gibbs, *Report on Physical Apparatus and Chemical Materials Suitable for Scientific Research* (Washington, D.C., 1876), p. 8.
[168] Edward S. Holden, ed., "Reports of Astronomical Observatories," *Annual Report . . . Smithsonian Institution 1880*, p. 636.

ROBERT MCKIM, donor of the observatory at De Pauw University (q.v.), built an observatory near his home in Madison, Indiana, in 1883. This observatory, reputedly the first in that state, was provided with two refracting telescopes made by Alvan Clark & Sons. One, of 4 inches aperture, had a portable equatorial mount. The other, of 6 inches aperture, had a fixed equatorial mount and various accessories made by Fauth.[168a]

JOEL HASTINGS METCALF (1866–1925), an astronomer connected with Harvard College Observatory, had a private observatory at Taunton, Massachusetts. Among his instruments was a Clark equatorial of 7 inches aperture.[169]

MIAMI UNIVERSITY, at Oxford, Ohio, bought the 12-inch Clark re fractor from Wesleyan University (q.v.) about 1920. They have since given it to the amateur astronomer Leslie Peltier (q.v.).

The apparatus used by ALBERT A. MICHELSON in 1879 for his first measurement of the speed of light incorporated five basic optical pieces. Sunlight, introduced by a heliostat, was reflected by a small revolving mirror to a lens of long focal length; a fixed mirror located at the focus of the lens reflected the light back through the lens to the revolving mirror, and the deflection of the light was measured by a micrometer. The heliostat was designed by Keith, and the micrometer was made by Grunow. The mechanical part of the revolving mirror was made by Fauth; the mirror itself, a disc of glass 1¼ inches across, silvered on the front surface, was made by the Clarks, as were the lens and the fixed mirror. Because of its unusually large focal ratio—reminiscent of 17th-century aerial telescopes, this lens was of 8 inches aperture and 150 feet focus—the lens was not achromatized. The fixed mirror, 7 inches wide, was actually one of the heliostat mirrors used in photographing the 1874 Transit of Venus (q.v.).[170]

In 1880, for the use of their students, the UNIVERSITY OF MICHIGAN purchased two instruments with Clark optics and Fauth mounts.[171] The

---

[168a] Edward S. Holden, ed., "Astronomy," *Smithsonian Institution . . . Annual Report . . . 1883*, p. 428.

[169] P. Stroobant, *Les observatoires astronomiques et les astronomes* (Brussels, 1907), p. 234.

[170] Albert A. Michelson, "Experimental Determination of the Velocity of Light," *Proc., American Association for the Advancement of Science*, vol. 28 (1879), pp. 130–135.

[171] *Publications, Astronomical Observatory of the University of Michigan*, vol. 1 (1912), history and apparatus. Also, private correspondence with Orren C. Mohler, Chairman, Department of Astronomy, University of Michigan.

3-inch aperture transit instrument, though inactive, is still in Ann Arbor. The 6-inch objective of the equatorial refractor is now used in the finding telescope of the main instrument in the Lamont-Hussey Observatory in South Africa.

The UNIVERSITY OF MISSISSIPPI, which ordered but was unable to purchase the Dearborn telescope (q.v.), finally received a sample of the Clarks' workmanship around 1868. At that time the Clarks reground the 5-inch objective of the university's G. Merz und Söhne equatorial refracting telescope.[172] Twenty years later, when the University of Mississippi was again in the market for a large telescope, Prof. R. B. Fulton asked the Clarks for a bid. Alvan Graham wrote in reply that they had been "giving much attention to medium sized telescopes calculated to do the finest possible work, and one that we have recently made for Harvard College, of 13 inches, is doing the most satisfactory work." The cost of a similar telescope, adaptable for photographic as well as visual work, and securely mounted, would be $10,000.[173] Despite the attractiveness of the Clark's offer, the contract was finally given to Sir Howard Grubb of Dublin.

In 1880 the UNIVERSITY OF MISSOURI, in exchange for their 1853 Fitz equatorial refractor of 4 inches aperture and $500, acquired the 7½-inch Merz und Söhne equatorial refractor which had been erected at Shelby College in Shelbyville, Kentucky, in 1850. At the time of transfer the larger telescope was guided by a Clark finder of 1⅞ inches aperture.[174]

In 1858 the "Women of America," acting through Elizabeth Peabody, gave MARIA MITCHELL, America's first woman astronomer, an equatorial refracting telescope made by the Clarks.[175] Miss Mitchell economized by taking a 5-inch objective, rather than one of 6 inches, so she

---

[172] List of Articles of Philosophical Apparatus in the Collection, 1861, with Later Additions (manuscript copy in Department of Physics and Astronomy, University of Mississippi).

[173] Alvan Graham Clark to Prof. R. B. Fulton, 18 August 1890 (letter in Mississippi Collection, University of Mississippi Library).

[174] Milton Updegraff, "Determinations of the Latitude, Longitude and Height Above Sea Level of the Laws Observatory of the University of the State of Missouri," *Trans., Academy of Science of St. Louis*, vol. 6 (1894), p. 483. See also Edward S. Holden, ed., "Reports of Astronomical Observatories," op. cit., p. 635.

[175] Helen Wright, *Sweeper in the Sky* (New York, 1949), pp. 126-127.

could have a position micrometer as well.[176] She used this instrument for observations of comets, planets, and double stars, and made it available to the astronomy students at Vassar College (q.v.). It is now in the Maria Mitchell Observatory on Nantucket, along with the Clark telescopes given by Mrs. Abbey and Miss Loines (qq.v.).

Another Clark instrument known only from one slight 19th-century reference is a $6\frac{1}{4}$-inch refracting telescope owned by MOORE of Lynn (Massachusetts ?) by 1877.[177]

The MORRISON OBSERVATORY was founded in 1875 and, until 1922, was affiliated with the Pritchett School Institute of Glasgow, Missouri; it has since been transferred to Central College at Fayette, Missouri, (qq.v.). The large equatorial refractor of this observatory was made by the Clarks.[178] This instrument, with an objective of $12\frac{1}{4}$ inches aperture and 17 feet focus, was provided with graduated hour and declination circles read by verniers, a clock drive regulated by a conical pendulum, and a filar micrometer of which either the field or the wires could be illuminated. Its cost, in the shop, was $6,000 in gold. With this telescope Alvan Graham discovered a close companion to the star 78 Pegasi;[179] in 1878 Carr Waller Pritchett and his son Henry discovered and studied the great Jovian red spot. When the telescope was completed George and Alvan Graham went west to mount it; and two years later George went again to Glasgow to help mount their new English transit instrument.

The Morrison Observatory has three other Clark instruments—a chronograph and two 4-inch aperture portable refractors. One of the small telescopes is dated 1880 and was acquired by the observatory at that time. It came complete with case for "ready and safe transportation," and was used especially on expeditions to observe transits and eclipses. It was furnished with a ring micrometer and comet and solar eyepieces, as well as a battery of common ones. Its mount, held on a heavy walnut tripod, could be adjusted for either equatorial or altazimuth

---

[176] Maria Mitchell to William C. Bond, 15 January 1859 (letter in Bond Papers, Harvard University Archives).

[177] "Size of the Principal Telescopes in the World," *Popular Science Monthly*, vol. 10 (1876-1877), p. 576.

[178] *Publications, Morrison Observatory*, vol. 1 (1887), pp. 5-8.

[179] Alvan Clark to Edward S. Holden, 8 March 1876 (letter in Lick Observatory Archives).

motions.[180] The other small telescope is dated 1882 but seems to have been a much later acquisition.

Most of the New England women's colleges have taught astronomy since their foundings, and MOUNT HOLYOKE SEMINARY at South Hadley, Massachusetts, is no exception. Charles A. Young, of Dartmouth and later Princeton, was giving lectures on astronomy to the female students as early as 1869; and when funds were offered in 1880 he helped plan the new observatory. The main equipment was a Fauth meridian circle and a Clark refracting telescope. The latter was equatorially mounted and provided with diffraction spectroscope, filar micrometer, and the usual finder, graduated circles, and clock drive. According to Young, the 8-inch objective "is almost entirely the work of the senior Alvan Clark, and is one of the most perfect specimens of his art."[181] This original objective, remounted in 1929, is still in use.

The Clark telescope replaced a simple 6-inch refractor which had been used at Mount Holyoke for many years. This smaller instrument was sent to a Huguenot college in South Africa, after the Clarks had repaired it and provided equipment necessary for its "enlarged usefulness."[182]

By 1896 the public HIGH SCHOOL IN NORTHAMPTON, Massachusetts, had a 4½-inch Clark refracting telescope, "quite well mounted and placed in an excellent dome."[183]

Since 1887 the Dearborn observatory and the 18½-inch refracting telescope (q.v.) have been established at NORTHWESTERN UNIVERSITY, in Evanston, Illinois.

OHIO STATE UNIVERSITY bought a portable 4-inch Clark equatorial in time for observations of the 1882 Transit of Venus.[184] Long neglected in favor of larger instruments, this telescope has recently turned up, under dust and cobwebs, and been placed in active service.[185]

---

[180] *Publications, Morrison Observatory*, vol. 1 (1887), p. 13.
[181] Edward S. Holden, ed., "Reports of Astronomical Observatories," op. cit., p. 663.
[182] Harriet E. Sessions, "The Study of Astronomy at Mount Holyoke Seminary," *Mount Holyoke Alumnae Quarterly*, vol. 3 (1919), pp. 17-19.
[183] *Popular Astronomy*, vol. 4 (1896-1897), p. 453.
[184] *Sidereal Messenger*, vol. 1 (1882-1883), p. 265.
[185] Private correspondence with Bruce C. Harding, Archivist, Ohio State University.

WILLIAM TYLER OLCOTT, the author of several popular books on astronomy, used a 4-inch aperture Clark refractor made in 1893. A wooden tripod supported the brass with nickel tube and a hand-driven worm wheel. Olcott later gave the telescope to Phoebe Haas (q.v.), who then gave it to the American Association of Variable Star Observers (q.v.), which in turn loans it to its members. The Olcott instrument is now being used by Walter Scott Houston.

A 6-inch Clark equatorial furnished with all necessary accessories—i.e., driving clock, finely divided circles, filar micrometer, etc.,—was part of the equipment of the observatory given to the UNIVERSITY OF THE PACIFIC by Charles Goodall (q.v.) and David Jacks in 1885.[186]

ROWLEY PATTERSON of Dansville, New York, had a 5-inch aperture Clark refractor dated 1881.[187]

In 1959 LESLIE C. PELTIER, an amateur astronomer in Delphos, Ohio, acquired the 12-inch equatorial which the Clarks had built for Wesleyan University (q.v.) in 1868, and which had been in use at Miami University (q.v.).[188]

In 1912 the Allegheny Observatory (q.v.), founded by a group of Pittsburgh citizens, was incorporated into the UNIVERSITY OF PITTSBURGH.

The National Observatory of PRAGUE, presently affiliated with the Charles University in the suburb of Ondřejov,[189] was planned and built, toward the end of the 19th century, by Josef and Jan Frič. These Bohemian "workmen-mechanicians" purchased their 8-inch refractor from the Clarks.[190]

In 1877 the Clarks helped equip the newly erected John C. Green Observatory for student instruction at PRINCETON UNIVERSITY. The unusual and highly successful telescope was undoubtedly designed in collaboration with Charles A. Young, who had just been appointed to the Prince-

---

[186] *University of the Pacific Catalogue* (1901–1902), p. 17.

[187] William H. Knight, "Some Telescopes in the United States," op. cit., pp. 394–395.

[188] Leslie C. Peltier, *Starlight Nights* (New York, 1965), p. 220.

[189] P. Stroobant, *Les observatoires astronomiques et les astronomes* (Tournai/Paris, 1931), p. 187.

[190] *Astronomie v Československu od dob Nejstarších do Dnoška* (Prague, 1952), p. 210. See also P. Stroobant (Brussels, 1907), p. 169.

ton faculty. Its 9½-inch objective was constructed on the Gaussian curves; and the distance between the components could be adjusted to give the best color correction for the work at hand, whether visual, spectroscopic, or photographic.[191] This lens combination never failed to show Young objects usually considered tests for a 12-inch,[192] and was said to exhibit less outstanding color than one of the ordinary form. The cost of this equatorial refractor was $3,000. For use with this telescope the Clarks also made a 3-cylinder chronograph and a spectroscope which could hold either a single prism or a diffraction grating.[193] This spectroscope,[194] together with a diffraction grating ruled in February 1880 by D. C. Chapman on a Lewis M. Rutherfurd engine, has recently been donated to the Smithsonian Institution (q.v.).

The 23-inch equatorial refractor of the Halsted Observatory at Princeton, installed in 1882, was a more typical Clark instrument. The bi-convex crown lens was separated from the bi-concave flint lens by over 7 inches, a distance sufficient to prevent the formation of "ghost" images. This separation also hastened the equalization of temperature between the lenses and the external air, as well as permitting an effective aperture one-half inch larger than otherwise possible with that flint component. The glass discs, cast by Feil, were more transparent and freer from bubbles and striae than were the large Chance discs of the U.S. Naval Observatory and University of Virginia instruments (qq.v.). The mount, although similar to that of the U.S. Naval Observatory 26-inch, was heavier and freer of vibrations. The tailpieces of the telescope, and of its 5-inch finder, were made the same size as that of the 9½-inch equatorial, so that the eyepieces and various auxiliary apparatus could be used with all three instruments.[195] This telescope was remounted in 1935 and has recently been transferred to the Naval Observatory for use in the Southwest.

---

[191] Edward S. Holden, "Astronomy," *Annual Record of Science and Industry* (1877), p. 47.

[192] Charles A. Young, "Measures of the Polar and Equatorial Diameters of Mars," *Observatory*, vol. 3 (1879), p. 471.

[193] Edward S. Holden, "Astronomy," *Annual Record of Science and Industry* (1877), p. 28.

[194] Charles A. Young, *The Sun* (New York, 1896), pp. 66–69, fig. 18 (p. 69).

[195] Charles A. Young, "The Twenty-three Inch Telescope of the Halsted Observatory at Princeton," *Proc., American Association for the Advancement of Science*, vol. 31, (1882), pp. 112–116.

In 1882 the Clarks made a long focus—about 8 feet—lens, and perhaps the mechanical parts as well, for a Littrow-type spectroscope for C. F. Brackett of the Princeton physical laboratory. This economical and convenient instrument used the same lens and tube for both the collimator and the viewing telescope. By a wise choice of curves, and the addition of a small central screen, the usual problem of diffuse light reflected from the lens surfaces was sufficiently reduced that the spectroscope was, in Brackett's opinion, equal to any of the ordinary, two-telescope construction.[196]

Carr Waller Pritchett, founder and director of the PRITCHETT SCHOOL INSTITUTE in Glasgow, Missouri, searched for many years for support for an astronomical observatory. Around 1867 he made plans for an observatory to be erected on the institute grounds, and asked the Clarks to make an 8-inch equatorial refractor. This instrument was actually completed before "unforeseen difficulties" forced Pritchett to cancel the order.[197]

In 1874 Pritchett found a reliable patron, in the person of Miss Berenice Morrison, and was able to build and equip the Morrison Observatory (q.v.) and use it in conjunction with his school. When the Pritchett School Institute was closed in 1922, the observatory was transferred to Central College (q.v.)

The third Clark objective larger than any yet made was mounted by the Repsolds for the Russian observatory at PULKOWA. In honor of the excellence of this instrument Alvan Clark and Georg Repsold received gold medals from Czar Alexander III.

Otto Struve had visited Cambridgeport in the summer of 1879 and, in spite of the modesty of their shop, he contracted with the Clarks for the 30-inch lens. He also insisted on paying an extra $1000—a figure George Clark thought much too high—for a rough equatorial mount for testing the objective; this mount later served for testing the Lick lens.[198]

The achievement of the Pulkowa lens inspired many expressions of American national pride. Just twenty years previously, during the Civil War, J. M. Gillis had struggled to equip the navy with American-made

---

[196] C. F. Brackett, "Note on the Littrow Form of Spectroscope," *American Journal of Science*, vol. 24 (1882), pp. 60–61.

[197] *Publications, Morrison Observatory*, vol. 1 (1887), p. 1.

[198] *Zum 50-Jährigen besteh̄n der Nicolai-Hauptsternwarte* (St. Petersburg, 1889), pp. 25–38.

optical instruments. And now Europeans were looking to the new world for the optical parts of their greatest astronomical telescope.

The 30-inch telescope was in use until World War II, when Pulkowa was destroyed by bombs. The large objective and several of the smaller observatory instruments were, fortunately, moved to safety. However, although the observatory has been rebuilt according to the original plans, the Clark object glass has not yet been remounted.

QUEEN'S UNIVERSITY in Kingston, Ontario, Canada, inherited the Kingston observatory (q.v.) and its equipment, including the 6¼-inch Clark equatorial refracting telescope of 1855.[199]

By 1893 RANDOLPH-MACON COLLEGE, at Ashland, Virginia, had a small astronomical observatory equipped with a refracting telescope, a reflecting telescope, a transit instrument and a sextant. The refractor, of 5¼ inches aperture, had been made by John Bryne and "worked over" by the Clarks.[199a]

By 1888 the magnificent RAYMOND HOTEL in Pasadena, perhaps for the entertainment of its guests, had a 4-inch Clark telescope.[200]

A 6-inch Clark equatorial, intended primarily as an adjunct to classroom instruction, was given to the UNIVERSITY OF ROCHESTER in 1876.[201]

CHARLES H. ROCKWELL, of Tarrytown, New York, observed the 1882 Transit of Venus with a Clark telescope of 6¼ inches aperture, provided with a full solar prism.[202]

RUGBY SCHOOL is the present owner of the 8¼-inch equatorial the Clarks made for Dawes (q.v.) in 1859–61. Although their evening observing time is greatly curtailed by school regulations, Rugby scholars use the telescope daily for solar studies. The telescope was given to them by J. M. Wilson, a maths master, who had bought it from Lowe (q.v.) in 1871. Except for a new electric drive to replace the faulty Cooke drive, the telescope has not been changed.[203]

---

[199] *Queen's University and College Calendar* (Session 1863-1864).
[199a] *Catalogue, Randolph-Macon College, 1893-94* (Richmond, Va., n.d.), p. 24.
[200] Edward S. Holden, *Handbook of the Lick Observatory* (San Francisco, 1888), p. 125.
[201] *University of Rochester Catalogue* (1876-1877), p. 30.
[202] *Sidereal Messenger*, vol. 1 (1883), p. 264.
[203] P. M. Kenrick, "Astronomy at Rugby School," *Hermes* (1966), pp. 58-60.

LEWIS MORRIS RUTHERFURD'S first photographic experiments were inspired by those at Harvard (q.v.). Late in 1857, therefore, he applied to the Clarks for a new clock drive for his equatorial. This new clock—which had a remontoir escapement similar to that of Bond's spring governor—was of the "highest merit." Several years later, using an 11¼-inch lens which he had focused for the photographic rays, supported on the Fitz equatorial mount and driven by the Clark clock, Rutherfurd, in New York City, was able to photograph ninth-magnitude stars.

In 1860 Rutherfurd prepared a telescope and camera for the U.S. Coast Survey Expedition to observe the solar eclipse from Labrador. The two components of the 4¼-inch objective, which were made by the Clarks, were separated so that the best visual and photographic foci were united.[204]

In 1941 the UNIVERSITY OF SANTA CLARA, in California, bought a 16-inch equatorial refracting telescope from the Pacific Electric Railway Company. This instrument, which is still in use, had been made by the Clarks in 1882 for Lewis Swift and had been taken by Swift to the Lowe Observatory (qq.v.).[205]

FRANK EVANS SEAGRAVE acquired in 1875 an 8-inch equatorial refractor made and mounted by the Clarks. For greater rigidity the telescope tube was patterned after the tubes often used for transit instruments: it consisted of two cones of riveted sheet steel. The position circles were twice graduated: the fine scale was read by verniers and a larger scale was painted on the edges of the circles for convenience in locating celestial objects. Among the various accessories of the telescope was a Clark position micrometer.[206] The telescope was housed in Seagrave's private observatory located first in Providence, Rhode Island, and later in the suburb of North Scituate.[207] The Seagrave observatory is now operated under the aegis of Skyscrapers, Inc.

---

[204] Lewis M. Rutherfurd, "Astronomical Photography," *American Journal of Science*, vol. 39 (1865), pp. 304–309.

[205] *Proc., Astronomical Society of the Pacific*, vol. 53 (1941), p. 349.

[206] Edward S. Holden, "Astronomy," *Annual Record of Science and Industry* (1878), pp. 37–38.

[207] Charles H. Smiley, "Frank Evans Seagrave," *Popular Astronomy*, vol. 42 (1934), pp. 1–2.

FIGURE 20.—Interior of the Dartmouth College Observatory showing the apparatus made by Alvan Clark & Sons for Charles A. Young. The large twice-traversed prism spectroscope is attached to the eye-end of the 9.4-inch aperture equatorial refracting telescope. Courtes Dartmouth College Archives.

In 1869 or 1870 JUSTUS M. SILLIMAN, then a student pursuing the chemical course at the Rensselaer Polytechnic Institute, examined the flame of the Bessemer process with a spectroscope made by Alvan Clark. This instrument, a common laboratory type, had a collimator, a viewing telescope, and a third tube carrying a graduated scale; the spectrum was formed by an equiangular flint glass prism.[207a]

Just as Charles A. Young had been advisor to Mount Holyoke College (q.v.), so David Todd of Amherst supervised the construction and outfitting of an astronomical observatory for the women of SMITH COLLEGE in Northampton, Massachusetts. Among the instruments he chose, and which were installed in 1886, was an equatorial refracting telescope with an 11-inch aperture Clark lens and a Warner & Swasey mount.[208]

Since their retirement from active service several Clark and Clark-modified instruments have found their way into the SMITHSONIAN INSTITUTION—in the collections of the Division of Physical Sciences of the United States National Museum—where they are being preserved and are available for study and exhibition. Of the Transit of Venus apparatus the museum has acquired the 7-inch plane mirror (cat. #327,709), the 5-inch achromatic objective (cat. #327,710), and the jaw micrometer (cat. #327,708) used on Kerguelen Island. From the U.S. Naval Observatory the museum has received the Merz und Mahler comet seeker (cat. #327,700) and the 9.6-inch Merz und Mahler objective (cat. #327,703), both of which the Clarks refigured. Vassar College has given the museum the 12-inch equatorial refractor made originally by Fitz and remade by the Clarks and Warner & Swasey (cat. #323,566). The original Warner & Swasey equatorial mount, from the telescope of Beloit College, is also in the museum (cat. #316,100). The most recent addition to the Smithsonian's collection is the spectroscope made by the Clarks in 1877 for Princeton University (cat. #328,611) (see fig. 23, p. 96).

The Jesuit STONYHURST COLLEGE, near Clitheroe, East Lancashire, England, has the Clark equatorially-mounted refracting telescope which

---

[207a] Justus M. Silliman, "On the Examination of the Bessemer Flame with Colored Glass and with the Spectroscope," *American Journal of Science*, vol. 50 (1870), pp. 301–302.

[208] William C. Winlock, "Astronomy for 1886," *Annual Report . . Smithsonian Institution 1887*, p. 155.

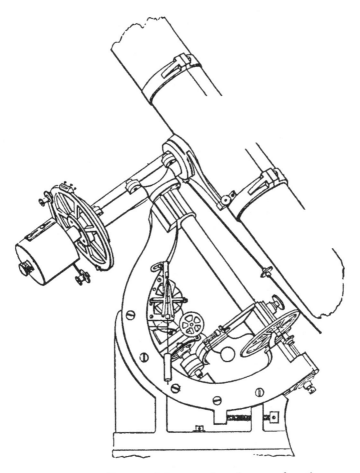

FIGURE 21.—Equatorial mounting for a refracting telescope, adjustable for use at any latitude, made by Alvan Clark & Sons, and described by William R. Dawes in the *Monthly Notices* of the Royal Astronomical Society. From *Monthly Notices,* vol. 20 (1859–1860), p. 62.

had belonged originally to Thomas William Webb (q.v.). The college authorities purchased the telescope in 1886, in preparation for the solar eclipse of that year.[208a] Soon thereafter the 5½-inch aperture Clark objective was remounted and used in conjunction with a heliostat to feed

---

[208a] *Stonyhurst College Observatory, Results of Meteorological, Magnetical and Solar Observations* (1886), pp. 6–7.

sunlight to a grating spectrometer inside the Stonyhurst College Observatory.[208b]

The student observatory at SWARTHMORE COLLEGE in Pennsylvania, built in 1886, houses a 6-inch equatorial refracting telescope complete with spectroscope and position micrometer, as well as a small transit instrument. The 6-inch objective was made by the Clarks, its mounting was by Warner & Swasey, and the accessories were made by Brashear.[209]

LEWIS SWIFT was the proud owner of a 16-inch Clark telescope—a present from his fellow citizens of Rochester, New York. When installed in the observatory built by H. H. Warner in 1882, it was the third largest refractor in the United States. It was equatorially mounted and, according to Swift, provided with all the modern improvements. These included a spectroscope and a filar micrometer with Burnham's illumination.[210] Aided by his son Edward, Swift searched for comets and nebulae until Warner went bankrupt and observing conditions in Rochester deteriorated badly. In 1893 the Swifts and their telescope moved to Mount Lowe (q.v.).

Swift's early observations had been made with a 4½-inch Fitz refractor; when the objective broke, around 1879, it was replaced by a Clark lens.[211]

The Holden Observatory at SYRACUSE UNIVERSITY, dedicated in 1887, has an 8-inch aperture Clark equatorial refractor[212]

Around 1882 Mr. C. W. TALLMAN of Batavia, New York, spent $400 for a 5-inch Clark refracting telescope with eyepieces magnifying from 25 to 500 times.[213]

ELIHU THOMSON, the electrical engineer of Lynn, Massachusetts, bought a 9-inch aperture Petzval photographic lens from a female photographer of Lynn for $70. In 1892 Alvan Clark & Sons informed Thomson that the lens "was made by us about 40 years ago. The surfaces

---

[208b] Ibid. (1888), p. 5.
[209] *Swarthmore College Catalogue* (1885–1886), p. 11.
[210] Lewis, Swift, *The History and Work of the Warner Observatory*, vol. 1 (1883–1886) pp. 7–10.
[211] Edward S. Holden, ed., "Reports of Astronomical Observatories," *Annual Report . . Smithsonian Institution . . 1880*, p. 660.
[212] *Sidereal Messenger*, vol. 7 (1888), p. 41.
[213] *Sidereal Messenger*, vol. 2 (1883), p. 23.

were polished on pitch. . . . We have never used cloth polishers."[214] If indeed this doublet was made in the early 1850's, it was the largest lens the Clarks had then made, and, except for Whipple's lens (q.v.), it is the only known photographic objective made by the Clarks at that time.

Each of the eight American parties authorized by Congress to observe the 1874 TRANSIT OF VENUS was outfitted with photographic apparatus, a 5-inch equatorial refractor, a chronograph—all made by the Clarks— and a Stackpole transit instrument with Clark optics, and two chronometers.[215] Appropriations for the equipment were not made until the summer of 1872, just two years before the first parties were to leave for their Pacific stations. George Clark apparently drove himself so hard to finish these instruments in time that, when they were finally delivered, he experienced a severe physical breakdown from which he never fully recovered.[216]

The various national expeditions used different methods for photographing the transit. The American system was an adaptation of Winlock's Harvard photoheliograph (q.v.). A heliostat directed sunlight onto a stationary photographic objective. The focal ratio of this objective— 5 inches aperture, and almost 39 feet focus—was so great that sufficiently large pictures could be taken in the focal plane, with no additional enlarging lens. The heliostat mirror was moved by a simple and inexpensive clock drive which needed only occasional manual adjustment. The mirror itself was of unsilvered glass, 7 inches in diameter, and slightly thicker on one side so that light reflected from the rear surface would be thrown away from the camera. The success of the photographs depended, to a great degree, on the perfection of the plane mirrors. The front surface of each mirror was to have a radius of curvature of not less than 4 miles; measurements made after the transit showed the mirrors to have more than twice the requisite flatness.

The purpose of the photographs was to determine the distance, at various times, that Venus had passed over the solar surface. To ensure precision, three special features were included: the photographic tele

---

[214] Alvan Clark & Sons to Elihu Thompson [sic], 25 November 1892 (letter in Thomson Papers, American Philosophical Society).

[215] Simon Newcomb, ed., *Observations of the Transit of Venus, December 8-9, 1874*, (Washington, D.C., 1880), pp. 14-16, 25-31, 61-65.

[216] "George Bassett Clark," *Proc., American Academy of Arts and Sciences*, vol. 27 (1891-1892), p. 362.

FIGURE 22.—Interior of the Lowe Observatory housing the 16-inch aperture equatorial refracting telescope built by Alvan Clark & Sons for Lewis Swift. Courtesy Southern Pacific Company.

scope was placed exactly in the meridian; a fine silver plumb line and a glass plate ruled with vertical and horizontal lines were placed immediately in front of the photographic plate; and the distance between the objective and the plate was measured as nearly as possible. Although none of the observations yielded a satisfactory value for the solar parallax, the American photographic method was deemed the most successful. The equipment was used again for the Transit of Venus of 1882, and for many other astronomical expeditions outfitted by the U.S. Naval Observatory. Some pieces of this equipment, only recently retired, have been given to the Smithsonian Institution (q.v.).

The 5-inch visual refracting telescopes were used for observing contacts and occultations of stars by the moon. They were equatorially mounted, and furnished with divided circles, clock drive, and double-image micrometer. Like the telescope the Clarks had made for Dawes in 1859 (q.v.), these were adjustable for any latitude. The chronographs were regulated by a Hipp spring, supposedly more reliable in the field than the Bond spring governor. These 8 telescopes, like the photographic apparatus, were for many years taken on astronomical expeditions around the world. In 1878, for instance, Alvan Graham used a Dallmeyer portrait lens, held on one of the Transit of Venus equatorial mounts, to photograph a total solar eclipse (see above, p. 33).

ETIENNE LEOPOLD TROUVELOT, whose astronomical observatory was located in Cambridge, Massachusetts, took good advantage of his neighbors. In preparation for his manual of astronomical drawings Trouvelot made observations with the U.S. Naval Observatory 26-inch telescope (q.v.), as well as with the 26-inch for the University of Virginia (q.v.), while they were still in the Cambridgeport workshop; and he frequently compared observations with Alvan Graham. Although without a Clark telescope of his own, Trouvelot did possess a Clark spectroscope with a diffraction grating ruled by Lewis Morris Rutherford.[217] During the 1878 solar eclipse Trouvelot used a small Merz refractor with a solar eyepiece and Barlow lens made by the Clarks.[218]

The U.S. ARMY BATTALION OF ENGINEERS operated an astronomical observatory in conjunction with their Engineer School of Application at

---

[217] Etienne L. Trouvelot, *The Trouvelot Astronomical Drawings Manual* (New York 1882), pp. vi, 77.

[218] *U.S. Naval Observatory Reports of the Total Solar Eclipses of July 29, 1878, and January 11, 1880* (Washington, D.C., 1880), p. 76.

FIGURE 23.—Spectroscope made by Alvan Clark & Sons in 1877 for use with the 9-inch refracting telescope at Princeton University. As shown here, the spectrum is formed by a Rutherfurd diffraction grating, but the instrument was also equipped with a glass prism for this purpose. The spectroscope is now in the collections of the Smithsonian Institution. The drawing is from Charles A. Young, *The Sun* (New York, 1896), fig. 18, p. 69.

Willets Point, New York Harbor. In 1880 they acquired an equatorial refracting telescope made by Fauth & Co., with a 5½-inch objective lens figured by the Clarks.[218a]

The U.S. MILITARY ACADEMY at West Point, New York, acquired a Fitz equatorial refractor of 9¾ inches aperture in 1856; in 1875 the Clarks reworked the objective of this instrument at a cost of $500.[219] In 1883 the West Shore Railroad, which had run a tunnel directly under the old observatory, built a new observatory for the cadets. The following

---

[218a] J. H. Willard, *Practical Astronomy at the Engineer School of Application at Willets Point, N. Y. H., 1881* (Willets Point, 1882).

[219] Alvan Clark to Peter S. Michie, 1 September 1875 (letter in U.S. Military Academy Library).

year a 12-inch equatorial refractor was purchased from the Clarks for $6280.[220] The clockwork of this telescope was regulated by the spring governor taken from the 12-inch formerly owned by Henry Draper, before it was sent to the Lick Observatory (qq.v.). The observatory and large refractor were seldom used. The tube has been scrapped and the objective lost. (See fig. 25, p. 100.)

The well-equipped astronomical observatory at the U.S. NAVAL ACADEMY at Annapolis was used primarily for instruction.[221] Their large refractor was made by the Clarks in 1857. The 7¾-inch achromatic objective had a focal length of 9 feet 6 inches (f/15 was the focal ratio of most Clark lenses); the German style equatorial mount was supported on a cast iron pier; the driving clock was regulated by a Bond spring governor. During the total solar eclipse of August 1869—the first astronomical occasion on which Clark instruments were extensively used—the Naval Academy's telescope was taken to Des Moines, Iowa, and used photographically by Dr. Edward Curtis of the Surgeon General's Office.[222] The Academy's observatory has since been closed, and the telescope has disappeared.

The Civil War occasioned a change in the direction of the U.S. NAVAL OBSERVATORY—a change which brought new jobs to the Clarks. Matthew F. Maury, a southerner as well as an oceanographer, resigned his commission in 1861. His successor, James M. Gillis, improved the instruments and promoted the astronomical work of the observatory.

In 1862 the Clarks refigured the objectives of most of the Washington instruments, always with decided improvement both in achromatism and definition. To replace the stolen 3.9-inch objective of the Merz und Mahler comet seeker, Gillis asked both the Clarks and Henry Fitz for bids; although the Clarks charged 50 percent more, and required much more time than did Fitz, they got the job.[223] This object glass turned out so well the Clarks were asked to rework the 5.3-inch lens of the Ertel merid-

---

[220] F. S. Harlow, "The Observatory of the U.S. Military Academy at West Point," *Publications, Astronomical Society of the Pacific*, vol. 3 (1891), pp. 273–274.

[221] Charles André and A. Angot, *L'astronomie pratique et les observatoires en Europe et en Amérique*. Pt. 3: *Etats-Unis d'Amérique* (Paris 1877), p. 112.

[222] *U.S. Naval Observatory Reports on Observations of the Total Eclipse of the Sun, August 7, 1869*, p. 124.

[223] Alvan Clark & Sons to James M. Gillis, 21 April 1862; Henry Fitz to James M. Gillis, 18 April 1862 (letters in U.S. Naval Observatory Papers, National Archives, Record Group 78).

FIGURE 24.—Telescope pier and mount in the yard of the Alvan Clark & Sons establishment used for testing the 30-inch aperture objective lens for the Russian observatory at Pulkowa. From *Scientific American*, vol. 48 (1883), p. 207.

ian transit, the 4-inch object glass of the Troughton & Simms mural circle, and the 9.6-inch lens of the Merz und Mahler equatorial.[224] Twenty-five years later they provided this last named telescope with a photometer for observations of variable stars.[225] When the Naval Observatory was rebuilt, at the end of the century, the 9.6-inch was replaced by a telescope with a 12-inch Clark objective and a Saegmüller equatorial mount.[226] The comet seeker and 9.6-inch lens are now in the Smithsonian Institution (q.v.).

During the Civil War the Naval Observatory, the former Depot of Charts and Instruments, supplied the Navy with its navigational instruments. Gillis, reluctant to rely on foreign manufacturers, wrote to Ameri can artisans, encouraging "the successful and permanent establishment of American factories for all classes of instruments." [227] In particular, he requested the Clarks to make spyglasses, as many and as soon as possible. The Clarks sent him their first sample in June 1863; in their typical, unbusinesslike fashion, however, they neglected to include their names and the price. During the next two years the Clarks sold the Navy at least 165 spyglasses at prices ranging from $25.75 to $35.00 apiece.[228]

In the fall of 1864 Gillis asked the Clarks to undertake binoculars for viewing a large field under high magnification—as this was the only foreign instrument he was still compelled to purchase. Robert Tolles, of Canastota, New York, had supplied him with some, but not enough, binoculars that were better than those available from Europe. The Clarks' first pair, sent to Washington in February 1865, were rated "very creditable" and "far better" than those attempted by Henry Fitz.[229] As the war ended shortly thereafter, the Clarks probably made no more binoculars for the Navy.

---

[224] *Astronomical and Meteorological Observations Made at the United States Naval Observatory During the Year 1862*, pp. vii–x.

[225] *Sidereal Messenger*, vol. 5 (1886), p. 88.

[226] A. N. Skinner. "The United States Naval Observatory." *Science*, vol. 9 (1899). p. 12.

[227] James M. Gillis to Alvan Clark & Sons, 26 January 1865 (letter in U.S. Naval Observatory Papers, National Archives, Record Group 78).

[228] See correspondence between Alvan Clark & Sons and James M. Gillis, 1863–1865 (letters in U.S. Naval Observatory Papers, National Archives, Record Group 78).

[229] Ibid.

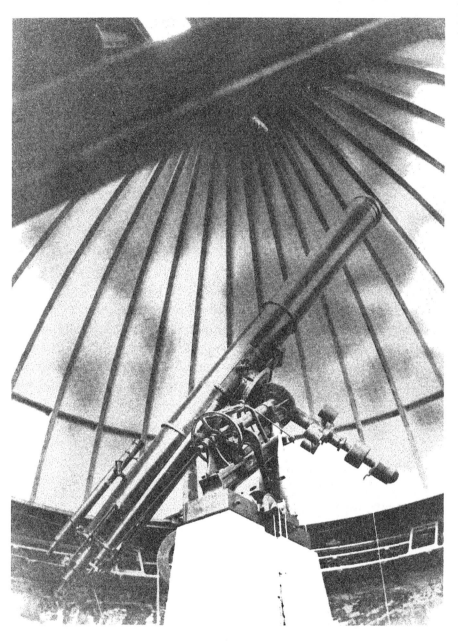

FIGURE 25.—The 12-inch aperture equatorial refracting telescope made by Alvan Clark & Sons for the U.S. Military Academy at West Point. Courtesy U.S. Military Academy Library.

Shortly after the installation of the Pistor & Martins meridian circle in 1865 the Clarks reworked most of its optical surfaces: the collimator lenses, the eyepieces, and the 8½-inch objective. When, along with all the other instruments, the meridian circle was renovated for removal to the new observatory, the Clarks made two new objectives for it. The first, ground on the Littrow curves, had excellent definition only near the optical axis. The second, and more successful glass, was of 9.14 inches aperture and had curves computed by William Harkness.[230]

As early as 1869 astronomers outfitted by the Naval Observatory used Clark portable equatorials for eclipse observations. These were usually of 3 inches apertures and 48 inches focus. Observers who were given, or used their own, instruments by other makers frequently kept their remarks about them to a minimum; observers using Clark telescopes, on the other hand, often named the opticians and praised their instruments.[231]

Although more inclined to precise measurement than to descriptive or astrophysical researches, the Naval Observatory did purchase several experimental, long focus, photographic telescopes from the Clarks.[232] Their first photographic apparatus, modeled after Winlock's equipment at Harvard (q.v.), was for use during the 1870 solar eclipse. Later ones helped determine the American method of observing the Transit of Venus (q.v.).

The great Washington equatorial—the 26-inch refractor of the U.S. Naval Observatory—was the second largest-ever Clark instrument, and the one which made the Clarks popular heros. The Dearborn refractor (q.v.) had been completed in 1863 and, owing to the war, received little public attention; however, the daily progress of the Washington telescope was publicly noted. The Dearborn instrument had been so thoughtlessly enclosed in a college tower that its potential could seldom be realized. Not so the Washington instrument; and the discovery of the satellites of Mars, during the opposition of 1877, demonstrated to all the power of this telescope.[233]

The Washington telescope was proposed with the Clarks in mind. While calling attention to their lack of a suitable instrument, the Wash-

---

[230] *Publications, United States Naval Observatory*, vol. 1, 2nd ser. (1900), pp. viii–ix.

[231] See *U.S. Naval Observatory Reports of Observations of the Total Eclipse of the Sun, August 7, 1869*, p. 27; and *U.S. Nautical Almanac Reports of Observations of the Total Eclipse of the Sun, August 7, 1869*, pp. 21–22.

[232] See correspondence between B. F. Sands and Alvan Clark & Sons, 1870–1872 (letters in U.S. Naval Observatory Papers, National Archives, Record Group 78).

[233] Simon Newcomb, *Reminiscences of an Astronomer* (Boston 1903), pp. 128–144.

ington astronomers emphasized Alvan Clark's competence, and ambition, to make the largest refracting telescope in the world [234]—that is, one larger, if only by half an inch, than the 25-inch equatorial recently built by Thomas Cooke for R. S. Newall in England.[235]

In America at that time there were neither glassmakers competent to cast the large discs nor other opticians on a par with the Clarks. Preparatory to beginning work on the Washington telescope, therefore, the Clarks made several European business trips. In September 1870 George went to Birmingham to engage Chance & Co. to cast the discs, and to Gateshead to see Newall's telescope. Four months later, after observing the solar eclipse from Spain, Alvan Graham went to England to examine Newall's instrument for himself and to see how Chance was progressing. In the fall of 1871 Alvan Graham returned to Birmingham to inspect the discs before shipment, and to the continent to study other large telescopes.

The discs were figured and tested by the fall of 1872; after a slight correction they were pronounced "very nearly perfect." The mounting was designed by the Clarks in collaboration with Simon Newcomb. Like the Newall telescope, the original Washington tube was made of sheet steel rather than wood. The driving clock was invented by Newcomb specifically for this instrument: it was driven by weights raised by water from the city pipes; and it was regulated by a conical pendulum whose isochronism was insured by an electromagnetically controlled friction pad. The Clarks were also responsible for several of the accessory instruments used with the great refractor: a two prism spectroscope, a chronograph regulated by a conical pendulum, a variety of eyepieces, and two micrometers.[236] The first micrometer was used until 1900, at which time the second, with the addition of an electric illumination, was installed. In his annual report for that year the superintendent of the observatory noted that this "new" micrometer, made by George Clark in 1874, "represents the finest workmanship of a gifted mechanic." [237]

---

[234] *Astronomical and Meteorological Observations Made at the United States Naval Observatory During the Year 1869*, p. v.

[235] Alvan Clark & Sons to Simon Newcomb, 17 June 1870 (letter in Newcomb Papers at the Library of Congress).

[236] *Instruments and Publications of the United States Naval Observatory* (Washington, D.C., 1876), pp. 26-45.

[237] *Report of the Superintendent of the United States Naval Observatory for the Fiscal Year Ending June 30, 1900*, p. 22.

FIGURE 26.—The 26-inch aperture equatorial refracting telescope made by Alvan Clark & Sons for the U.S. Naval Observatory in Washington, D.C.

The mounting of the 26-inch telescope lacked stability and so in 1893, when the observatory moved to its present location in Georgetown Heights, the great equatorial was remounted by Warner & Swasey. The discarded mounting and driving clock were salvaged and formed the basis of a photographic telescope. The tube, objective, and micrometer of the old 9.6-inch aperture refractor, mounted inside the shortened tube of the 26-inch equatorial, served as a guiding telescope. Photographs were taken with twin 6-inch Dallmeyer portrait cameras—like that used by Alvan Graham Clark to photograph the 1878 solar eclipse—mounted outside the large tube, and later with twin cameras with 10-inch wide lenses.[238] In 1921 the weight-driven drive of the photographic telescope was replaced by an electric one.[239]

Maria Mitchell, at VASSAR COLLEGE, called on the Clarks to make and repair her astronomical instruments. The 12½-inch Fitz equatorial refractor—which has recently been retired to the Smithsonian Institution (q.v.)—had a troublesome original objective and mount. In 1868, shortly after it was installed, the Clarks reground the lenses to such a degree that they considered this telescope one of theirs. Soon thereafter they repaired the driving apparatus.[240] Warner & Swasey later completely remounted the telescope.

The smaller equipment of the Vassar observatory included a meridian instrument by Young of Philadelphia, with two collimating telescopes by the Clarks,[241] and two portable Clark refractors of 3 and 6 inches aperture.[242] As early as 1871 Vassar students took pictures of the sun, every day at noon, with a 2-inch aperture, non-achromatic, long focus photographic telescope, similar to that used by Winlock at Harvard (q.v.), and made by the Clarks.[243] The Vassar students had the use of several other portable telescopes as well: the largest, a 5-inch Clark equatorial, belonged to Miss Mitchell (q.v.).

---

[238] George Henry Peters, "The Photographic Telescope of the U.S. Naval Observatory," *Popular Astronomy*, vol. 27 (1919), pp. 1–10.

[239] George Henry Peters, "The New Electric Driving Clock of the Photographic Telescope of the U.S. Naval Observatory," *Popular Astronomy*, vol. 30 (1922), pp. 1–10.

[240] Maria Mitchell, Observations with the 12-Inch Equatorial 13 April 1866–12 June 1877, entries for October 1868, 18 August 1870, and 7 January 1871 (Manuscript copy in Vassar College Library).

[241] Edward S. Holden, "Astronomy," *Annual Record of Science and Industry* (1877), p. 51.

[242] *The Vassar Transcript*, February 1879. See also Mary W. Whitney, "Vassar College Observatory," *Publications, Astronomical Society of the Pacific*, vol. 6 (1894), p. 151.

[243] Alvan Clark & Sons to Simon Newcomb, 31 May [1871?] (letter in Newcomb Papers at the Library of Congress).

FIGURE 27.—The Vassar College Observatory, with Maria Mitchell (1818–1889), left, and Mary W. Whitney (1847–1921).

In 1873 the Clarks sold a 12-inch achromatic objective to the Austrian government for the Imperial Observatory at VIENNA. As the lens was not delivered until the new observatory was built, Alvan Graham had the use of it for several years. During this time he discovered at least five pairs of double stars, including the companion of γ Lyrae.[244]

In 1871, while arrangements for the 26-inch equatorial of the U.S. Naval Observatory (q.v.) were being made, Leander McCormick ordered a similar refracting telescope for his native state of VIRGINIA. The Clarks would have completed this instrument within a few years, but held off for a while because of McCormick's financial difficulties. This delay gave them a chance to learn from some of the mistakes of the Washington instrument. The inner surfaces of the Virginia 26-inch objective, for instance, were given slightly different radii, so as to avoid the annoying "object-glass ghost." The driving clock was connected with a Seth Thomas clock located in the computing room of the observatory. The filar micrometer, like that at the University of Wisconsin (q.v.), was provided with Burnham's illumination. The telescope was finally erected at the Leander McCormick Observatory at the University of Virginia in 1884.[245] This observatory also boasts a 6-inch Clark equatorial refractor of 1892.[246]

Between 1882 and 1893 the WARNER OBSERVATORY in Rochester, New York, housed the 16-inch Clark equatorial refracting telescope belonging to Lewis Swift (q.v.).

Many of the telescopes originally made in the 1850's by Henry Fitz were later extensively remodeled. One such instrument is the 6½-inch refractor at WASHINGTON UNIVERSITY in St. Louis, Missouri. In 1882-83 the Clarks reground the objective and Warner & Swasey made a new mount.[247]

The physics department of WASHINGTON AND JEFFERSON COLLEGE, at Washington, Pennsylvania, now has possession of the 7½-inch objective

---

[244] S. W. Burnham, "Double Stars Discovered by Alvan G. Clark," *American Journal of Science*, vol. 17 (1879), p. 285.

[245] "Leander McCormick Observatory," *Scientific American*, vol. 60 (1889), p. 55.

[246] P. Stroobant, *Les observatoires astronomiques et les astronomes* (Tournai/Paris, 1931), p. 50.

[247] From private correspondence with Richard H. Lytle, University Archivist, Washington University.

originally made by the Clarks for the telescope of Jefferson College (q.v.)

THOMAS WILLIAM WEBB counted among his common telescopes a refractor with a Clark objective of 5½ inches aperture and 7 feet focus. Acting on the advice of W. R. Dawes (q.v.), Webb purchased this telescope in 1859 and used it for observations of the great comet two years later.[248] In 1886, after Webb's death, the telescope was sold to Stonyhurst College (q.v.).

WILLIAM HARVEY WELLS, as mentioned above (p. 19), was the Clarks' first telescope customer. His 5-inch refractor, mounted in 1848, showed the three stars of γ Andromeda and sometimes the sixth star in the trapezium of Orion.[249]

There are two equatorial refractors in the Whitin Observatory at WELLESLEY COLLEGE. The larger is the 12-inch Fitz/Clark telescope, formerly owned by Jacob Campbell (q.v.) and Stephen V. C. White (q.v.). Sarah Frances Whiting, Wellesley's first astronomer, had been envious of this telescope since her teaching days in Brooklyn when she had taken her students to look through it—through the finest glass in the vicinity.[250] The other Wellesley instrument is a 6-inch telescope marked at the eye-end "ALVAN CLARK & SONS CAMBRIDGEPORT, MASS. 1890"; the mount, which is probably of a later date, is marked "ALVAN CLARK & SONS CO."

OLIVER CLINTON WENDELL, a Harvard astronomer, had a 6½-inch Clark equatorial refractor in his Lowell observatory by 1878. He also had, on loan, a portable 3½-inch Clark refractor.[251] As none of their diaries or private correspondence has yet been found, it is difficult to discover much about the Clarks' personal lives. It is apparent, however, that Wendell was especially friendly with the Clarks. The obituary of Alvan Graham that Wendell wrote for *The Cambridge Tribune* was prepared at the request of the family, as one who stood near him. It,

---

[248] *Astronomische Nachrichten* # 1348, vol. 57 (1862), p. 61.

[249] Elias Loomis, *Recent Progress of Astronomy* (New York, 1850), p. 252.

[250] Annie J. Cannon, "Sarah Frances Whiting," *Popular Astronomy*, vol. 35 (1927), p. 4.

[251] Edward S. Holden, "Astronomy," *Annual Record of Science and Industry* (1878), p. 66.

as well as the various other biographies from Wendell's pen, shows a sympathetic awareness of the optician's personality.[252]

In 1868 WESLEYAN UNIVERSITY, at Middletown, Connecticut, built a second astronomical observatory and equipped it with Clark apparatus. Except for small pieces, Clark instruments were usually made to order. The surviving correspondence between the Clarks and Prof. John M. Van Vleck of Wesleyan shows that, in this case, the Clarks not only designed the instruments but gave advice on the construction of pier and dome as well.

The new dome enclosed a 12-inch refractor of about 15 feet focus; this was equatorially mounted and provided with finder, circles, driving clock, and micrometer. The cost, including delivery to the observatory, was $6,000 in gold.[253] After fifty years service this telescope was transferred to Miami University (q.v.). It has recently been given to Leslie Peltier (q.v.).

During the solar eclipse of 7 August 1869, Van Vleck used, for the first time, a Clark spectroscope attached to a 3½-inch portable Clark equatorial.[254] The spectroscope was designed for use either with the large equatorial or in the laboratory. It employed a single glass prism and a scale of equal parts for locating spectral lines. The Clarks apparently did not realize the importance of precise spectroscopic determinations. They made this instrument to answer all purposes except exact measures because, "the observations for locating the lines are excessively tedious and but few persons have the patience or time to do much with them." [255]

Van Vleck apparently intended to order a chronograph from the Clarks. In a letter of 1868 the Clarks replied that they had never made chronographs, that Wm. Bond & Son was the chief concern in that busi-

---

[252] Oliver C. Wendell, "A Tribute From Mr. Clark's Personal Friend," *The Cambridge Chronicle*, 26 June 1897. See also Oliver C. Wendell, "Alvan Graham Clark," Proc., *American Academy of Arts and Sciences*, vol. 33 (1897–1898), pp. 520–524; and Oliver C. Wendell, "Alvan Graham Clark," *Astrophysical Journal*, vol. 6 (1897), pp. 136–137.

[253] Alvan Clark & Sons to Prof. J. M. Van Vleck, 10 April 1868 (letter in Wesleyan University Archives).

[254] *U.S. Nautical Almanac Reports of Observations of the Total Eclipse of the Sun, August 7, 1869*, pp. 81, 87.

[255] Alvan Clark & Sons to Prof. J. M. Van Vleck, 18 November [1868?] (letter in Wesleyan University Archives).

ness.[256] By 1874, however, the Clarks had expanded their line and made chronographs for the American observers of the Transit of Venus (q.v.).

The only photographer's objective made by the Clarks that has yet been found—except for the Petzval doublet purchased by Elihu Thomson (q.v.)—belonged to JOHN A. WHIPPLE. In 1863 a visitor to Whipple's Boston studio reported that "His mammoth lens, the work of Alvan Clark of Cambridge, is a sight to see, as are also the views which it takes."[257] Although Whipple pioneered in celestial photography, there is no record that his Clark lens was used for anything other than studio portraits.

STEPHEN VAN CULEN WHITE bought Jacob Campbell's (q.v.) telescope, observatory, and house in Brooklyn. When the American Astronomical Society was founded in 1883, White, as the owner of the largest and finest refracting telescope of any private observer in America, was elected president.[258] Fifteen years later White's telescope was given to Wellesley College (q.v.) where it is still in use.

WILLIAMS COLLEGE, in Williamstown, Massachusetts, added an equatorial refracting telescope to the equipment of their Hopkins Observatory in 1852. The Clarks, who had yet to prove their mechanical skills, made the optical parts while Jonas Phelps of Troy, New York, made the mount. The 7⅛-inch achromatic objective, probably the largest the Clarks had yet made, was deemed excellent.[259] Prof. Albert Hopkins of Williams was so well pleased with the telescope that he recommended a similar one be purchased for Jefferson College (q.v.).[260] The Williams College telescope, unchanged except for a new electric motor drive, is still in use.

In 1890 the principal instrument in C. W. WILSON's private observatory in Lynn, Massachusetts, was a 6-inch Clark telescope of "unusual excellence."[261] This is very possibly the telescope which had formerly been owned by Moore (q.v.).

---

[256] Alvan Clark & Sons to Prof. J. M. Van Vleck, 10 April 1868 (ibid.).
[257] "Photography in Boston," *American Journal of Photography*, vol. 6 (1863–1864), p. 322.
[258] *Scientific American*, vol. 48 (1883), p. 64.
[259] Elias Loomis, *Recent Progress of Astronomy* (New York, 1856), p. 209.
[260] Albert Hopkins to William C. Bond, 7 February 1859 (letter in Bond Papers, Harvard University Archives).
[261] William C. Winlock, "Progress of Astronomy for 1889, 1890," *Annual Report Smithsonian Institution* . . 1890, p. 163.

The Washburn Observatory of the UNIVERSITY OF WISCONSIN is equipped with several Clark instruments.[262] Their equatorial refractor, like that of the U.S. Naval Observatory (q.v.), well illustrates the quantitative competitiveness of 19th-century observatories: the order for the telescope specified that it be "larger and more powerful" than the great refractor of the Harvard College Observatory (q.v.). The Wisconsin instrument, built in 1878, had an aperture of 15.56 inches and was provided with eyepieces of various powers, a polarizing solar eyepiece, a double-image micrometer, and a filar micrometer with bright-wire illumination. This last was designed by, and built for, S. W. Burnham, who was then observing difficult double stars—that is, pairs of unequal brightness, and pairs at the limit of the telescope's separating power. George Clark produced an elegant micrometer in which the wires were sharply defined and uniformly bright in all positions and which avoided stray light in the field. The micrometer proved so successful that similar ones were made for such instruments as the 26-inch refractor at the University of Virginia and the 12-inch at the Lick Observatory (qq.v.). The Clarks' equatorial mount for the Washburn telescope was not as successful as the micrometer and optical parts; within twenty years it was found less convenient than more modern instruments.[263]

The meridian circle at the University of Wisconsin, made by Repsold in 1881, has a 4.8-inch Clark objective. When S. W. Burnham (q.v.) went to Madison he took his 6-inch Clark refractor with him and sold it to the university in 1882.

The astronomical observatory of the Sheffield Scientific School at YALE UNIVERSITY was opened in 1866. Its equatorial refracting telescope, from the Clark workshop, was said to perform in accordance with the reputation of its makers. The achromatic objective had a clear aperture of 9 inches and a focus of only 9 feet 10 inches. The tube, which was of stiff but light pine, was one of the last wooden ones made by the Clarks. As in the instrument which had so impressed Dawes (q.v.), a U-shaped iron piece supported the polar axis and protected the driving

---

[262] *Publications, Washburn Observatory*, vol. 1 (1882), pp. 23-30. See also Joel Stebbins, "The Washburn Observatory, 1878-1958," *Publications, Astronomical Society of the Pacific*, vol. 70 (1958), p. 438.

[263] G. C. Comstock, "The Washburn Observatory," *Publications, Astronomical Society of the Pacific*, vol. 9 (1897), p. 32.

FIGURE 28.—Alvan Graham Clark and Carl A. R. Lundin with the crown glass component of the 40-inch aperture objective lens for the Yerkes Observatory. From George E. Hale, *Study of Stellar Evolution* (Chicago, 1908), plate xlv.

clock and spring governor of the Yale telescope.[264] Later Clark additions to the observatory included a portable refractor of 4⅔ inches aperture, a conical pendulum chronograph, a multiple ring micrometer, and a spectroscope of seven prisms twice traversed.[265]

The Winchester Observatory at Yale tried unsuccessfully to obtain a Clark instrument. Their Repsold heliometer, the first such instrument in the United States, arrived in time for use during the 1882 Transit of Venus. The optical work had been done by Merz only after Alvan Clark had declined the responsibility of dividing the 6-inch objective.[266] The

---

[264] J. E. Nourse, "Observatories in the United States," *Harpers New Monthly Magazine*, vol. 49 (1874), p. 530.
[265] Edward S. Holden, "Astronomy," *Annual Record of Science and Industry* (1878), pp. 68–69.
[266] A. M. Clerke, "The Yale College Measurement of the Pleiades," *Sidereal Messenger*, vol. 6 (1887), p. 254.

Winchester Observatory was also in the possession of several blank glass discs: a 29-inch diameter flint, and two flint and two crown pieces 10 inches wide. In 1875 these pieces were given to the Clarks for examination and safe storage. Extant correspondence indicates that, had money been available, the Yale astronomers would have asked the Clarks to figure a 29-inch achromatic objective.[267] Money to work the lenses, however, was not found during the Clarks' lifetimes.

The last objective made by the Clarks is at the YERKES OBSERVATORY of the University of Chicago. In respect to both the figure of the lens, and the form of the Warner & Swasey mount, the Yerkes telescope is similar to the Lick instrument (q.v.). Like the Dearborn telescope (q.v.), the 40-inch went to Chicago by default. While in California in 1888–89, Alvan Graham discussed the possibility of a large refractor for Wilson's Peak with representatives of the University of Southern California; and he received authorization to place an order with Mantois of Paris for the 40-inch crown and flint discs. Soon thereafter the Californians found themselves without the funds necessary for purchase of such a telescope. Although there were frequent notices in newspapers and scientific journals of the existence of these discs, and of the Clarks' desire to figure them, no definite bids were made until Alvan Graham met George Ellery Hale at the September 1892 meeting of the AAAS.[268] The persuasive Hale quickly talked C. T. Yerkes into paying for an observatory and a mount, as well as for the great objective.

The 40-inch lens, the largest one made by the Clarks, is as yet the largest one ever made and mounted. Had Alvan Graham lived longer— he died in June 1897, shortly after delivering the objective to the observatory at Williams Bay, Wisconsin—lenses of even greater aperture might have been made. The often mentioned problems of adequate support and absorption of light through the glass seemed trivial to him. In 1893, when visited by Edward Emerson Barnard, Alvan Graham expressed his readiness, as soon as the 40-inch was figured, to begin work on a 5-foot objective.[269]

---

[267] See H. A. Newton to Prof. Lyman, 12 August 1884 (letter in Yale University Archives).

[268] Edwin B. Frost, *An Astronomer's Life* (Boston, 1933), p. 97.

[269] Edward E. Barnard, "Nearer to the Stars," *English Mechanic*, vol. 60 (1894–1895), pp. 495–496.

Appendix

Paintings by Alvan Clark

Alvan Clark painted portraits and miniatures, both as an avocation and a vocation, for about 37 years. How many he painted is not known, but a good estimate would be around 500. As mentioned above (p. 7), in Boston alone, Clark earned over $20,000 painting "heads," and for each he probably charged around $40, the price paid by Lucius Manlius Sargent for his ivory miniature.

An attempt to identify the subjects of Clark's pictures has yielded, so far, only about eighty "heads." The task has been difficult as Clark seldom signed his paintings and seems to have been even more reluctant to date them. Furthermore, it is quite likely that families have held onto portraits, for sentimental if not aesthetic reasons, and relatively few Clark paintings have as yet found their way into museum collections.

The following is a list of the subjects of Clark paintings, together with medium, size and present and past owners, as far as they are known. Some are well identified, others are known only through brief mentions in articles on the Clark establishment or reminiscences by various Clark descendants.

In the compilation of this list I have been aided by the museums and private persons who have Clark paintings in their collections.

Francesca Alexander (probably daughter of the portrait painter, Francis Alexander, 1800–ca. 1881). Miniature of baby girl, owned by Mrs. Caroline Clark Eastman. Frick Art Reference Library (cited hereafter as FARL) #21,656.

David Andrews. Miniature, owned by Mrs. Helen N. Elderkin of Old Greenwich, Conn. FARL #9,549–B.

John Anthon (probably 1784–1863, lawyer). Miniature, exhibited 1831 at the National Academy of Design, New York, N.Y.

Caroline Bartlett. Watercolor on ivory, 4¾ x 3⅛ in. Museum of Fine Arts, Boston, Mass.; bequest of Minna B. Hall.

John Bass. Watercolor on cardboard, 7⅝ x 6¼ in. Museum of Fine Arts, Boston, Mass.; bequest of Grenville H. Norcross.

Adelaide Bass. Watercolor on cardboard, 7 x 5½ in. Museum of Fine Arts, Boston, Mass.; bequest of Grenville H. Norcross.

Maria Bass. Watercolor on cardboard, 7⅛ x 5¾ in. Museum of Fine Arts, Boston, Mass.; bequest of Grenville H. Norcross.

Mary Bass (Mrs. Stephen G. Bass). Watercolor on cardboard, 7⅛ x 5¾ in. Museum of Fine Arts, Boston, Mass.; bequest of Grenville H. Norcross.

Mary N. Bass. Watercolor on cardboard, 7⅛ x 5¾ in. Museum of Fine Arts, Boston, Mass.; bequest of Grenville H. Norcross.

Stephen G. Bass. Watercolor on cardboard, 7⅛ x 5¾ in. Museum of Fine Arts, Boston, Mass.; bequest of Grenville H. Norcross.

Elisha Bassett (1745–1832, maternal grandfather of Alvan Clark). Oil on board, 17½ x 10½ in. Loaned by Miss Cornelia Church to the Ashfield Historical Society Museum, Ashfield, Mass.

Francis Bassett. Portrait painted in middle age, showing grand head of fluffy white hair, ¾ view.

Electra Chamberlin Bement (Mrs. Jasper Bement). Watercolor on cardboard 3½ x 3 in. Museum of Fine Arts, Boston, Mass.; bequest of Annie Villette Bray and Elmer E. Bray.

Dr. Samuel A. Bemis (1793–1881, Boston dentist). Watercolor on ivory, 3 x 2½ in. Metropolitan Museum of Art, New York, N.Y.; Fletcher Fund. Originally owned by Dr. Bemis, who bequeathed it to Alvan Graham Clark; the Museum purchased it from his daughter, Caroline Clark Eastman.

Joseph Bishop, Esq. (of Albany, N.Y.). Oval miniature, fully signed and dated 1840 on reverse. Advertised for sale by E. Grosvener Paine, New York, N.Y., in 1966.

Ann Hill Blake. Miniature exhibited 1890 in Newport, R.I.

William Few Chrystie. Ink sketch, as a young man. Owned by William Few Chrystie; sold by Parke-Bernet Gallery, New York, N.Y., in 1949. FARL #36,795.

Constable Clapp. Portrait, hung for many years in the Alvan Clark & Sons workshop.

Clark family group—Alvan and Maria Pease Clark with three of their children, Alvan Graham, Caroline Amelia, and Maria Louisa. The lower right hand corner of this painting is blank, reputedly because Alvan Clark could not paint his own hands. Owned by Theodore C. Hollander, grandson of Alvan Graham Clark.

Clark family group. Alvan Clark is reputed to have painted several portraits of his wife and children.

Miss Clark (daughter of Alvan Clark, either Caroline Amelia or Maria Louisa). Portrait on wood panel, owned by Mrs. Harry Freeman, granddaughter of Alvan Graham Clark.

Abram Clark (1771–1835, father of Alvan Clark). Miniature, now owned by Mrs. Albert W. Rice of Worcester, Mass., inherited from her father, T. H. Gage.

Alvan Clark (1804–1887). Oil on canvas, 25 x 22½ in. The Cleveland Museum of Art, Cleveland, Ohio; gift of Francis H. Bigelow in memory of Lawrence Park.

Alvan Clark (1804–1887). Self-portrait, loaned by Mrs. Alvan Clark Eastman, granddaughter-in-law of Alvan Graham Clark, to the Museum of Fine Arts, Boston, Mass.

Alvan Clark (1804–1887). Oval miniature, owned by Theodore C. Hol lander, grandson of Alvan Graham Clark.

Alvan Clark, Jr. (1870–1884, only son of Alvan Graham Clark). Portrait, painted from photographs and memory after his death. Destroyed by fire.

Alvan Graham Clark (1832–1897). Two identical portraits, of different sizes, of a very young boy presumed to be Alvan Graham Clark. Owned by Theodore C. Hollander, grandson of Alvan Graham Clark.

Alvan Graham Clark (1832–1897). Portrait, as mature man.

Barnabus Clark (1799–?, oldest brother of Alvan Clark). Oil on canvas, 27⅜ x 22¼ in. National Gallery of Art, Washington, D.C.; Mellon Collection.

Barnabus Clark (1799–?, oldest brother of Alvan Clark). Miniature owned by Theodore C. Hollander, grandson of Alvan Graham Clark.

Barnabus Clark (1799–?, oldest brother of Alvan Clark). Watercolor on ivory, 2⅞ x 2⁵⁄₁₆ in. Worcester Art Museum, Worcester, Mass.; formerly owned by Elizabeth W. Grogan, daughter of Alvan Graham Clark.

Caroline Amelia Clark (1835–1863, daughter of Alvan Clark). Portrait.

George Bassett Clark (1827–1892). Miniature, as a baby asleep on his pillow.

Maria Pease Clark (1808–?, Mrs. Alvan Clark). Watercolor on ivory, 2³⁄₁₆ x 1¹³⁄₁₆ in. Metropolitan Museum of Art, New York, N.Y.; bequest of Glenn Tilley Morse, 1950.

Maria Pease Clark (1808–?, Mrs. Alvan Clark). Oval miniature, 3 x 2⅜ in. Owned by Theodore C. Hollander, grandson of Alvan Graham Clark.

Mary Bassett Clark (1773–1855, mother of Alvan Clark). Oval miniature, owned by Theodore C. Hollander, grandson of Alvan Graham Clark.

Mary Mitchell Willard Clark (1846–1892, wife of Alvan Graham Clark). Large oval portrait, owned by Theodore C. Hollander, grandson of Alvan Graham Clark.

Henry Codman. Oil on panel, 12¼ x 10 in. Museum of Fine Arts, Boston, Mass.; bequest of Maxim Karolik.

Mrs. Henry Codman. Oil on panel, 9½ x 8½ in. Museum of Fine Arts, Boston, Mass.; bequest of Maxim Karolik.

Stephen Codman. Watercolor on ivory, 3¼ x 2⅝ in. Museum of Fine Arts, Boston, Mass.; bequest of Maxim Karolik.

Charles Henry Cummings. Oil on canvas, 27 x 22 in. Museum of Fine Arts, Boston, Mass.; gift of Miss Mabel Cummings.

Mrs. Charles Henry Cummings. Oil on canvas, 27 x 22 in. Museum of Fine Arts, Boston, Mass.; gift of Miss Mabel Cummings.

Walter Forward (1783–1852, statesman). Miniature, in Carnegie Museum, Pittsburgh, Pa.

Mr. Frothingham (probably James Frothingham, 1786–1864, artist). Portrait, copied from picture by H. (J.) Frothingham, exhibited 1830 at the Boston Athenaeum.

Mrs. Mary Caroline Goddard. Oil on ivory, 3 x 2½ in. Philadelphia Museum of Art, Philadelphia, Pa.; gift of Mrs. Daniel J. McCarthy.

Chester Harding (1792-1866, artist). Oval miniature, exhibited 1838 and 1839 by the Apollo Association at Harding's Gallery, Boston, Mass., and at the National Academy of Design, New York, N.Y., in 1838. Owned by Theodore C. Hollander, grandson of Alvan Graham Clark.

Robert Hare (1781-1858, chemist). Oil on canvas, 25 x 30 in. National Portrait Gallery, Smithsonian Institution, Washington, D.C.

Caroline Virginia Harris. Oval miniature, girl aged about 9, dressed in white. Owned by D. D. Hamlen; sold by Parke-Bernet Gallery, New York, N.Y., in 1946. FARL #36,873.

Dr. Thomas Hill (1818-1891, president of Harvard College). Portrait, now owned by a descendent.

Mrs. David Murray Hoffman. Miniature, owned by William Wickham Hoffman of New York, N.Y. FARL #50,631.

Mrs. H. B. Humphrey. Miniature, exhibited 1890 in Newport, R.I.

Anna Maria Levy (1805-1899, Mrs. David Cardoza Levy). Watercolor on ivory, oval, 1⅞ x 2 9/16 in. Signed "A.C." lower left. Maryland Historical Society, Baltimore, Md.; Eleanor S. Cohen Collection.

David Cardoza Levy (1805-1887). Watercolor on ivory, oval, 2 x 1⅝ in. Signed "A.C." lower left. Maryland Historical Society, Baltimore, Md.· Eleanor S. Cohen Collection.

George Livermore (1809-1865, author, antiquarian). Watercolor on ivory, 2 5/16 x 2 7/16 in. Unsigned and undated. Yale University Art Gallery, New Haven, Conn.; formerly owned by Mrs. John Hill Morgan.

Nathan Loomis (mathematician, astronomer). Oil on canvas, 27 x 22 in. Unsigned and undated. Owned by the Clark family until the turn of the century, when it was transferred to a descendent of Prof. Loomis· now owned by Mrs. Millicent Todd Bingham, granddaughter of Loomis. This painting has been loaned to the Yale University Art Gallery.

Elizabeth Salisbury Lovedell. Painting, owned by M. Irving Motte of Concord, Mass. FARL #17,988.

Gilman Low. Oil on canvas, 36 x 29 in. Museum of Fine Arts, Boston, Mass.; gift of Mrs. Walter R. Eaton.

Dr. Mitchell. Miniature, exhibited 1839 by Apollo Association at Harding's Gallery, Boston, Mass.

Joseph Addison Norcross. Watercolor on cardboard, 7 x 5¾ in. Museum of Fine Arts, Boston, Mass.; bequest of Grenville H. Norcross.

W. Page (probably William Page, 1811–1885, artist). Miniature, exhibited 1838 and 1839 by Apollo Association at Harding's Gallery, Boston, Mass.

Beula Pease (probably sister of Maria Pease Clark). Miniature, owned by Mrs. Caroline Clark Eastman, daughter of Alvan Graham Clark. FARL #21,655.

Joseph Ripley. Miniature, owned by Mrs. Helen N. Elderkin of Old Greenwich, Conn. FARL #5,974–A.

Mrs. Joseph Ripley. Miniature, owned by Mrs. Helen N. Elderkin of Old Greenwich, Conn. FARL #5,974–B.

Mrs. Joseph Ripley. Miniature, owned by Mrs. Helen N. Elderkin of Old Greenwich, Conn. FARL #5,974–C.

Lucius Manlius Sargent (1786–1867, temperance advocate). Ivory miniature, painted in Fall River, Mass., in 1835, for which he paid $40.

Benjamin Shattuck. Owned by Miss Bessie Howard.

Mrs. Benjamin Shattuck. Owned by Miss Bessie Howard.

Miss Shattuck. Owned by Mrs. Archibald Taylor. FARL #45,878.

Mrs. Henry Smith. Miniature, exhibited 1836 at the Museum of Fine Arts, Boston, Mass.

Ammi Burnham Stiles (1815–1877). Oil on canvas, 28½ x 23 in. Essex Institute, Salem, Mass.

Hanna Annis Stiles (1817–1904, Mrs. Ammi Burnham Stiles). Oil on canvas, 28½ x 23 in. Essex Institute, Salem, Mass.

Joseph Story (1779–1845, judge). Portrait, exhibited 1846 at the Boston Athenaeum.

Mrs. R. H. Stuart (cousin of Alvan Clark). Wash drawing, done while Alvan Clark was itinerant artist in the Connecticut Valley. Owned by Mrs. Stuart's son.

Mrs. R. H. Stuart (cousin of Alvan Clark). Portrait, painted in later life. Owned by Mrs. Stuart's son.

Mr. Taft. Oval miniature. In the American Art Association's Herbert Lawton Collection; sold at the Anderson Gallery, 1937. FARL #30,512.

Joseph Tryon. Watercolor on cardboard, 6 x 4½ in. Museum of Fine Arts, Boston, Mass.; bequest of Grenville H. Norcross.

Mrs. Joseph Tryon. Watercolor on cardboard, 6 x 4½ in. Museum of Fine Arts, Boston, Mass.; bequest of Grenville H. Norcross.

George Washington (1732–1799, general, statesman). Oil on canvas, 36 x 30 in. From the original by Gilbert Stuart in the Boston Athenaeum. Unsigned. U.S. Naval Academy Museum, Annapolis, Md.; transferred 1869 from the Navy Department.

Daniel Webster (1782–1852, statesman). Portrait, owned by the Clark family.

Mrs. Mark Wentworth (1822–1848, nee Susan Osgood Jones). Portrait, owned by S. Wentworth of Wilton, N.H.

Lovice Corbett Whittemore (Mrs. Thomas Whittemore). Oil on canvas, 30 x 24 in. National Gallery of Art, Washington, D.C.; gift of Thomas Whittemore, 1950.

Thomas Whittemore (1800–1861, clergyman). Oil on canvas, 30⅛ x 25⅛ in. National Gallery of Art, Washington, D.C.; gift of Thomas Whittemore, 1950.

John Greenleaf Whittier (1807–1892, poet). Watercolor on ivory, oval, 3 x 2½ in. Museum of Art, Rhode Island School of Design, Providence, R.I.

Unidentified man–1 (man with red hair). Watercolor on ivory, oval, 3 x 2⅝ in. Unsigned and undated. National Collection of Fine Arts, Smithsonian Institution, Washington, D.C.; purchased by Tallman from Mrs. C. V. Wheeler in 1948; Wheeler purchased it from Brooks, Reed Gallery, Boston, Mass., in 1916.

Unidentified man–2. Oval miniature, owned by Theodore C. Hollander grandson of Alvan Graham Clark.

Unidentified man–3. Oval miniature, owned by Theodore C. Hollander, grandson of Alvan Graham Clark.

Unidentified man–4. Watercolor on ivory, 4 x 3 in. Loaned by Alvan Clark Eastman, grandson of Alvan Graham Clark, to the Museum of

Fine Arts, Boston, Mass. This is probably the miniature listed in the Frick Art Reference Library (FARL #21,657) as belonging to Caroline Amelia Clark Eastman, mother of Alvan Clark Eastman.

Unidentified woman–1 (lady in a blue dress). Watercolor on ivory 3⅜ x 2⁹⁄₁₆ in. Unsigned. Metropolitan Museum of Art, New York, N.Y.· Fletcher Fund. Formerly owned by Herbert Lawton, and then by Erskine Hewitt.

Unidentified woman–2. Oval miniature, owned by Theodore C. Hollander, grandson of Alvan Graham Clark.

Unidentified. Miniature, owned by Mrs. Harry Freeman, granddaughter of Alvan Graham Clark.

Made in the USA
Monee, IL
31 March 2023